> " *All advanced technology is initially*
>
> *indistinguishable from magic.* "
>
> (Arthur C Clarke)

I

The secret is out!

It's easy to release clean, green, zero-carbon hydrogen from water for the sustainable future of our planet!

One day we will surely look back at our use of crude oil and ask "What on earth were we thinking of?" It's so easy to release hydrogen from water efficiently, but they didn't want you to know that as there's little profit in it. In fact, oil is mainly carbon and hydrogen and it's the hydrogen that is volatile. Carbon is much harder to ignite and is the source of most of our health and environmental problems. Engines don't work well with carbon either. Water, being made of hydrogen and oxygen, is made of two volatile gases and only needs a small current to release them. That's not what they taught you at school, is it?

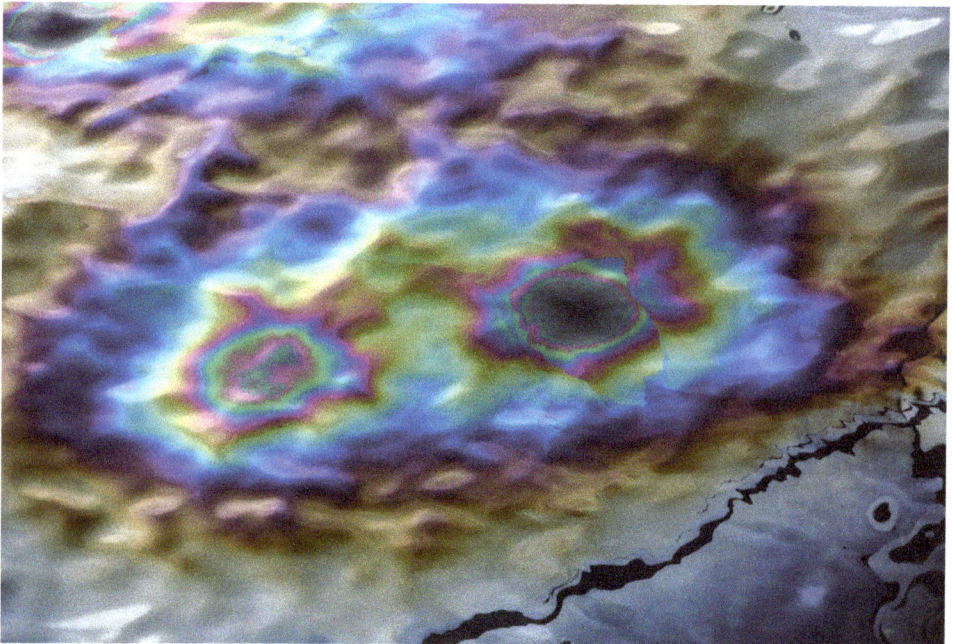

No Carbon Required!

Oxyhydrogen from Water, the Perfect Fuel of the Future

First published in 2022 by Paul Adams

Time is running out, the whole planet is in crisis, and oil is too precious to burn. The technology outlined in this book has been until now suppressed by greed. I urge you, the reader, to consider what is outlined herein and help to promote a sustainable future of benefit to all.

All it takes is a RESONANT FREQUENCY to efficiently release HYDROGEN on-demand from WATER

It doesn't break the Laws of Physics It updates and redefines them!

...SO WHY HAVE THE PIONEERS OF THIS TECHNOLOGY BEEN DISBELIEVED, BRIBED, OUTLAWED, RIDICULED, SILENCED, EXILED AND EVEN MURDERED?

If anything should happen to me, as it did to people like pioneer Stanley Meyer (chapter 11), this time it wouldn't prevent this technology getting out to the wider world. In fact it would have the opposite effect. I would become a martyr, my mission complete and the world asking questions.

Why? Because, not only is the book now in major circulation, but also I have copied and downloaded all the referenced material, such as patents and web archives and videos and spread copies of all my research to interested parties as well as the hugely secure Web3 blockchain (see p271). My entire body of research is open-source.

No Carbon Required

Hydrogen-on-demand from plain water

by Paul Adams

The Pentagon knows about this proven technology, NASA knows about it, the American Navy knows about it, Rolls Royce aerospace also knows about it. Covering it up is the crime of the century, resulting in the Climate crisis, irreparable damage to the environment, damage to our health, untold economic misery for everyone struggling to pay for electricity for our homes, fuel for our vehicles and power for everything from construction to farming. It's even caused wars while dealing with rogue oil-rich nations. Now is the time to reveal this 'secret' to unlocking the power hidden within water, although, I fear, it may already be too late.

Sourcing fuel from water involves many disciplines: Physics, chemistry, geology, geography, mechanical engineering, electronics, electrical engineering, computing, music theory and even politics.

Note about web links

Many of the reference links in this book have been shortened for your convenience, however, they can all be found on the website www.hydod.com as direct links. Please let me know if any are broken or redundant.

In this book, I'm going to reveal to you a novel technique using *resonance* to split water into usable hydrogen and oxygen that can then be used to *replace all fossil fuels in any engine.* This would, of course, mean a *carbon-free* future and revolutionise the way the world is powered and have *massive economical implications* too as there are many free sources of water, particularly seawater. This is a technology that isn't particularly complex or expensive, but has been hidden from public view as using plain water (or seawater), would mean *the end of big oil profits* and the *vast government taxes* that go with them. Although it might sound unfeasible, *it has been scientifically proven* (If you need scientific proof before reading any further, see page 252 and 253), and several people including a retired NASA engineer, have used it to power generators, motorcycles, cars, motorhomes and boats, all of which I shall give references to later in the book, including links to Google patents, YouTube videos and other sources of information that you can research yourself. I'll reveal how these genius innovators have always been intimidated and quickly shut down by the authorities.

All truth passes through three stages:

First, it is ridiculed;

Second, it is violently opposed;

Third, it is accepted as self-evident.

Arthur Schopenhauer (1788-1860)

It works!

At least a dozen people have successfully run their engines on water over the years, including motorcycles, cars, boats, and generators (See chapter 17). There are also patents freely available for all to see (many of which have expired). In this book I'll introduce you to the most important of those, looking at them in some detail and what background information you need to understand exactly how it works. I'll cover what the molecular makeup of water is, and how it can be successfully fractured using a resonant frequency, as well as the properties of hydrogen, one of its constituents, and carbon, its nemesis. Also, I'll introduce you to the basics of the scientific principles that are involved, without going into too much technical detail or using complex technical jargon. I'll also cover why it is so important to bring this technology to the fore and look at how it will be of benefit to everyone on the planet to switch to this abundant source of clean, zero-carbon, nonpolluting energy.

Forward by Dick Bell MBE

(Former RAF Test Pilot for the Lightning Jet Aircraft).

KILLER CARBON

We All need carbon – we are part made of it.

BUT NOT THIS MUCH.
Look what it's doing. It's killing us all.

We all love our cars, trains, aeroplanes, buses and ships.
We all love our nylon, plastic and other kinds of man-made materials.

BUT NOT THIS MUCH.

ALL THAT STUFF IS MADE OF FOSSIL FUELS – petrol, diesel, coal, natural gas. We heat our homes, live comfy lives, chuck all our plastics away without a care. We drive to our supermarkets instead of walking. We kill our oceans, our trees and forests and soon our planet – and kill mankind (you and me) as collateral.

Fossil fuels are brim full of carbon. Tons of it. Every time we burn it we pour nasty, neat, dirty, black carbon into our nice clean air, so we don't have any nice clean air anymore.

Remember the lock-down for Covid-19? Everyone stayed at home so no one drove. Indians in Mumbai could see the Himalayas, Parisians in Paris could see the Eiffel Tower, Brits in London could see St Paul's (at long last). Smog had gone. Covid came and then went, and no one can see anything anymore, and some school children walking to school by main roads died of asthma.

WE HAVE GOT TO STOP BURNING FOSSIL FUELS.

But, we meekly ask, is there an alternative? Surely, "Qué será, será"? ("Whatever will be, will be").

No, No, No. THERE IS AN ALTERNATIVE. It's not inevitable. Our grandchildren do not need to die at age 25.

Clean, cheap. Available fuel in mammoth quantities. Easy to make available. What is this incredible alternative?

VIII

Read the following as a matter of extreme urgency. It is a matter of Life and Death. SAVE OUR PLANET. And please, people, stop greedy oil magnates hiding this truth from the rest of us. They are making millions to line their fat pockets – all from continuing the lie that there is no alternative to flooding the world with more and more black, gungy, filthy, polluting oil.

WONDERFUL HYDROGEN.

What is the alternative? HYDROGEN.

And here is the fantastic bit – Hydrogen from **WATER**.

We've got plenty of water, 71% of Earth's surface is covered in it. Water can be provided throughout every continent, especially in those places where water is scarce today, particularly if we spend less on digging up very expensive fossil oil, gas and coal, and use the same technology to drill for water.

Water? Yes. Water is made up of two beautiful gases – Hydrogen and Oxygen, H_2O. Hydrogen burns sweetly. It leaves no pollution, gives off great heat, is instantly ignited, and is tucked away in water. Oxygen is lovely stuff, and makes every fire we know burn better. (That's why the wildfires we see on Earth today burn fiercer with a wind). Besides, we breathe Oxygen. It is beautiful stuff. We can also get loads of that from water.

All we have to do is to separate the Hydrogen and the Oxygen from our water, and Hey-Ho! we've got plenty of fuel for our cars, our trains, our ships, our aeroplanes, our homes, and all our mechanical moving parts. With NO pollution. (Put the Himalayas, the Eiffel Tower and St Paul's back on the agenda again).

Is it easy to do? YES, YES, YES.

We don't have to have expensive technology like we were all taught at school. Lots of clever people have devised *much* cheaper and simpler ways of extracting the Hydrogen and Oxygen from water. It has all been patented for decades. A guy in the USA built a car that ran on it. Another drove an RV thousands of miles right across America on water alone. You don't need huge storage tanks. We only need to change the water into hydrogen and oxygen two inches from our car's engine intake, and we can run it for far longer than we can on nasty fossil-fuel carbon-polluting petrol or diesel. And we can fill our fuel tanks up with water! Think of the savings! Wouldn't that be wonderful?

We can heat our homes with hydrogen too. And run our huge electricity power stations on it. Every internal combustion engine can be easily converted to burn hydrogen. The more we do it, the cheaper the conversion becomes. And think about aeroplanes. Filling their tanks up with less water than fossil fuel will be a fantastic safety feature, and stop polluting our skies. And what about ships? They could sail in oceans of available fuel.

All the research is done. The science is complete.

NOW, PEOPLE OF PLANET EARTH, GO AND DO IT.

THIS BOOK TELLS US HOW.

READ IT. THEN TELL EVERYBODY ELSE TO READ IT, AND WE CAN CHANGE THIS PLANET IN ONLY A FEW YEARS, IF WE ALL CLUB TOGETHER AND PUT OUR MINDS AND INDUSTRIES TO IT.

CHEAPLY.

BE GONE, GLOBAL WARMING!

THIS BOOK IS MORE VALUABLE THAN GOLD OR PINK DIAMONDS, BECAUSE *ALL* OF US WILL BENEFIT FROM ITS WISDOM.

Dick Bell MBE

Doubters will doubt!

Be very aware of claiming that something is 'Not Possible'

For example:

"X-rays will prove to be a hoax"

Lord kelvin,

"There is not the slightest indication that nuclear energy will ever be obtainable. It would mean that the atom would have to be shattered at will."

Albert Einstein, 1934 in the Pittsburgh Post-Gazette

"I can state flatly that heavier than air flying machines are impossible."

Lord kelvin, President of the Royal Society, Acclaimed mathematical physicist, 1895,
(8 years before a pair of bicycle engineers proved him spectacularly wrong).

What this book is about.

Well, what is the biggest problem in the world right now? Climate change. And what is the cause? The burning of fossil fuels. Is it really possible to split water into hydrogen and oxygen efficiently using electricity to replace all fossil fuels completely?

After many years of research and a career as an insider in the electric power industry as well as coal and oil industries as a technical author, I can tell you that it is. This book is my attempt to convince the world to hopefully stop using carbon-rich fuel immediately.

I am writing this during the Coronavirus lockdown of Spring 2020. During this time, whilst everyone kept themselves safely in their homes around the world, something amazing was happening to the planet. As only essential travel was allowed, pollution caused by traffic fumes and industry was down. There were no jet contrails in the big blue sky. In the big cities notoriously polluted, such as Mumbai, in India, the sky was so clear that some people could see the distant Himalayas for the first time ever.

Many millions of people die annually of lung-related problems due to the combustion of fossil fuels in internal combustion engines of our transport, from cars to boats, trains and planes, as well as in commercial vehicles, farm vehicles, industrial engines, and power plants used to produce electricity. It doesn't have to be that way. There are alternatives, but historically governments and the big oil companies have done their utmost to suppress any viable alternatives until now.

There has been a lot of talk recently about hydrogen power and electrically powered vehicles but both of these have their problems. Hydrogen is very volatile and therefore difficult and dangerous to store or transport. Electric vehicles have charging issues, such as the need for infrastructure to be in place and the need for massive heavy and expensive batteries to store the power. However, there is an alternative and that is to produce hydrogen on demand by extracting it from plain water.

At this point, scientists and technical people will most likely leave the room and run as fast as they can in the opposite direction. They will claim that this idea is simply not possible due to the 'first law of thermodynamics' which states that you can only get out the energy that you put into the system. (Re: Principle of conservation of energy: 'energy can be transformed from one form to another, but can be neither created nor destroyed'). However, that depends on how efficiently you are applying the energy and what you consider to be part of the system. With oil, for example, it is accepted that refined oil is combustible when it is extracted from the ground, but water, although it is also similarly composed of highly combustible hydrogen gas, for some reason, isn't considered to be combustible.

In virtually all science textbooks and school and college and university labs, it is taught through a process called electrolysis, which introduces a direct current of electricity through water. You can split the water molecule into its component parts of hydrogen and oxygen but only on a small and not at all useful scale. What I'm about to tell you in this book is the story of how it is possible to produce a useful quantity of hydrogen with very little input of electricity. In other words a large amount of hydrogen with very little current. How is this possible? The use of an alternating current at a specific frequency which will weaken the water molecule sufficiently to make it easy to break the molecular bonds and extract the gas.

In this book, I hope to convince you that it not only is possible, but has been done on several occasions but continues to be suppressed by governments and greedy oil companies. It is so important that now we begin to understand the possibilities of using one of the world's most common resources which will produce *zero emissions* for all our transportation needs and the possibility of generating electricity and meeting all our requirements for industrial power as well. For the sake of the planet as well as the health of humanity, we must consider these possibilities now as a matter of great urgency. In the Epilogue, I will tell you what I propose to do about it and how, if you wish, you can get involved. The new science I'm introducing here has proved the viability of creating hydrogen on demand from water, but we need to convince the scientific community beyond any doubt, plus, there is a correct procedure to follow to do that. I am hopeful that this will now be possible.

Contents

Prologue:

"It ALWAYS seems impossible until it's done"

Nelson Mandella

I want this book to be for everyone, but due to its subject matter, it is, of course, unavoidable that many parts will be technical in nature. However, I have deliberately kept it as simple as possible in the body of the work, referencing more in-depth explanations in the Appendices. More detailed technical explanations can be found in documents available to download or referenced online from the hydod.com website. A simple internet search will clarify many of the processes but bear in mind that some of the principles are not conventional and some people may be inclined to consider them pseudoscience (For example, in some of the terminology used by pioneers such as Stanley Meyer and Andrija Puharić struggling to explain less understood phenomena at play). I, therefore, urge you to look closely at the facts, before leaping to conclusions. I am a great believer in science, but it is unwise to think that everything is understood by modern scientific knowledge. Do you want to progress, or be stuck in the 19th Century?

A note about the technical terminology used in this book.

If you are not of a technical background, you may find that a lot of the material in this book goes over your head. But please do not despair. If you look again you might realise that many of the technical terms are self-explanatory. For example, talk of a 'resonant frequency' may sound unfamiliar, but we often talk about 'resonating' with someone or something. And 'frequency' means an oscillating sound, just like a musical note. So a 'resonant frequency' just means a sound wave (or electrical wave) that 'resonates' with something else, making it vibrate in unison. Frequency is also just a measure of how often something happens, such as how many buses per hour, or how often per minute your heart beats. The latter can be clearly seen as a *waveform* on a hospital screen.

'Inductance 'and 'capacitance' are terms often used in the explanations too. 'Inductance' is when an electrical current in a wire, 'induces' a similar current in another one close by. 'Capacitance' is simply the capacity of something, like the capacity of a vessel like a jar or container. Here it is often used when talking about 'capacitors' which hold electric charges as batteries do. A 'transformer' literally 'transforms' one current into another, to a different (higher or lower) value. A 'field', is just that, an area similar to a 'playing field' or 'football field'. Here it's used for a whole area around a wire or perhaps a metal bar that contains an electric charge or magnetic charge. So please, if you will, try to understand a little more without shying away from the apparent complexity. It's not as complex as you might think!

My mission here is to bring an understanding of this technology to a wider audience. That could also be youngsters who are our future scientists and technicians. Many scientific 'facts' are perhaps just assumptions and theories, which hold until a new one comes along or the old one is disproved. Always keep an open mind.

In 1889, Charles H. Duell was the Commissioner of the US patent office. He is widely quoted as having stated that the patent office would soon shrink in size, and eventually close, because… "Everything that can be invented has been invented."

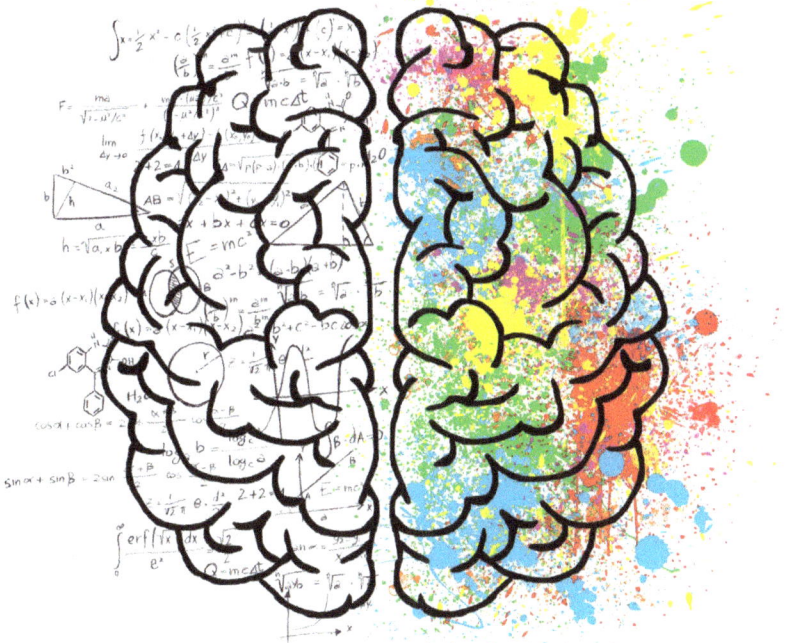

Always keep an open mind.
The logical left side and the creative right-hand side of the brain.
Most people tend to favour one side or the other.
Perfect coordination of the two hemispheres leads to genius.

Scientists and technical people tend to be 'left brained' people and artists, musicians, designers, innovators and inventors (and dyslexics), all right-brained people. Perhaps that is why radical ideas are often slow to be adopted.

Introduction

There's a lot of talk about climate change, every time you turn on the TV or look at a news website these days. It's also often proposed that hydrogen is possibly the solution to the world's energy problems. Solar, wind and wave power are reliant on fickle weather and economies of scale. The harsh environments are not kind to the infrastructure either, making maintenance very expensive.

Electric vehicles seem to be all the rage too but are not so environmentally friendly as they are charged with electricity still mainly provided by coal- or gas-fired power stations or nuclear power stations which have their own environmental hazards. Batteries are very expensive and there are issues surrounding the range of the vehicle and how and where to charge them. It can take hours to charge an electric vehicle, which is highly inconvenient, plus, the infrastructure is not growing anywhere near the rate of electric vehicle sales, posing future problems. And although there are rapid charging points, it's not generally good for the battery life span.

Hydrogen can be a totally clean, green energy, but there are some major hurdles to overcome that nobody seems to know how to solve. (Currently, it is mostly produced by steam reforming from natural gas - a fossil fuel!) For example, hydrogen is extremely volatile, very difficult to contain and store and dangerous to transport. There is however a solution to all this, namely to *produce the hydrogen on-demand at source*, then <u>no storage would be necessary</u>.

Plain water essentially consists of hydrogen and oxygen, bonded with shared electrons. All the schools, colleges and universities in the world, teach that production of hydrogen the conventional way, by splitting water (electrolysis, see chapter 8), is simply not efficient, requiring more energy input than you get from the resulting hydrogen. (The first law of thermodynamics often used as an argument to dispute the viability of sourcing hydrogen from water).

However, the traditional technique uses DC electricity at high current and low voltage. To succeed, what is required is **AC electricity** with a fluctuating voltage at a very precise **resonant frequency** (matched to both the water and fuel cell) and **very low current**, therefore making it extremely efficient and economical. Rapidly pulsing the voltage breaks the molecular bonds by flexing the water molecule in stages using a 'gated' pulse, until it breaks, thereby releasing the hydrogen and oxygen. It's similar to the way an opera singer can break a glass by hitting the correct note which precisely matches the natural note of the glass.

Several people have researched, experimented and achieved this with water, but have been disbelieved at best or outlawed by the big oil corporations who do not want such a freely available resource as water being used as fuel, not to mention governments not being able to extract tax revenue from fuel sales.

What changes things is the current climate emergency due to burning fossil fuels, full of carbon, coupled with the fact that oil reserves are dwindling. This also means that governments have carbon-zero targets to meet within stringent time scales.

One major hurdle to this new way of obtaining hydrogen on-demand from water is that the science is on the fringe of our understanding of how electricity and magnetism work. That means, unconventional terminology and techniques. In this book. I hope to clarify all these issues and convince the most sceptical people, especially scientific and technical people, *who often refuse to take a second look.*

After researching this thoroughly for several years, I am convinced that it is the solution to many problems. It's what the world is waiting for.

My intention, indeed my mission, is to throw this knowledge out into the world, in the hope that it will grow virally so that many people can start to understand and hopefully begin to use it. This time, because I have published this book, the secret is already out there as we speak, plus I am making details open-source via the website, which means freely sharing the ideas with everybody (Most of the patents I refer to are out of date too).

Stanley Meyer, designed and built a hydrogen-powered water-fuelled car and paid with his life (see chapter 11). I believe that his mistake was to try to go it alone. His mission was similar to mine. He saw that it would benefit the world and refused bribes of $1 billion from the Arab oil producers intended to prevent him from taking it forward. (The fact that it worried them so much speaks volumes!) He didn't accept the bribe as he realised the huge benefits to the world. However, he still thought he was going to do it all himself. He paid the ultimate price for that. The powers that be very nearly succeeded in suppressing it for good. This time they won't be able to.

Hydrogen on-demand from water is good to go

So how soon can we use this technology?

The big advantage of using water as a source of hydrogen, is the cost and availability, as well as the fact that there is no requirement for hydrogen storage at all.

Hydrogen has already been used in cars, buses, trucks and even aircraft, plus I read that it's possible to just use existing infrastructure, such as gas pipes for hydrogen instead of natural gas (which, after all, is just hydrogen and carbon). A coal-fired power station has already been successfully retrofitted to use hydrogen and being a gas similar to LPG (liquid petroleum gas), hydrogen can also be used in many applications, from cooking and heating properties in remote locations and campsites, to field kitchens as well as vehicles, from fork-lift trucks to cars, motor homes (RV's), vans and buses. I'm not sure if it's possible to directly exchange LPG for hydrogen, but both being gases, I'm sure it won't take many adjustments to make it work.

Hydrogen can be used to fuel generators too, with millions of uses from hospital power backups to remote electricity generation anywhere that the need arises. They can pump water and provide valuable power for use in emergency off-grid situations. Desalination is a problem currently, due to the high cost of fuel required to process large volumes of water at scale. If you can use water as a source for that, then "problem solved!". For industrial applications, hydrogen is superior to other fuels and gases, as it burns with an extremely pure flame. For cutting and welding, it has been proven to be superior to oxygen-acetylene torches. For applications in domestic situations, such as heating and air-conditioning, it has the advantage that it is completely non-toxic and nonpolluting. There have been several times that people have sadly died due to carbon monoxide poisoning from systems that use fossil fuels, not to mention explosions.

As we shall see in later chapters, there is only a need to produce a resonant frequency to split the water molecule efficiently. That can be provided by an alternator or even an audio signal generator and amplifier (such as that proposed by A. Puharić, (see chapter 10)). So, in short, this technology of on-demand hydrogen with water as a source can be used to produce electricity for heating and ventilation systems, for cooking, pumping water, desalination, industrial use including iron foundries, welding and other industrial uses, farm machinery, and every kind of transport imaginable. And all entirely CARBON FREE! So yes, it's good to go and as soon as possible.

The billions currently being thrown at 'low carbon' solutions and 'renewables' can surely be reassigned to this technology. Organisations with better resources and budgets than mine should start exploring all the scientific principles that I have outlined with great urgency. The government's lost tax revenue can be offset by actually achieving climate targets that they themselves have set. They will also benefit from the massively reduced cost of running government vehicles, heating and air-conditioning their buildings, as well as local and global travel. Zero carbon emissions by 2030? It suddenly all becomes possible.

Looking a bit closer

If we are going to split the water molecule, we must first take a much closer look at what water actually is and its properties, which as it turns out are rather odd. We first need to look at the molecular structure. Molecules are made from atoms, the basic building blocks of all physical matter in the universe. Atoms are like tiny planetary systems with electrons orbiting the nucleus similar in a way to the earth orbiting the sun, or the moon orbiting the earth. If we gave the moon a little push, a little more energy, it would speed up, resulting in it going into a 'higher' orbit. If we applied enough energy we could knock it off its orbit altogether and it may sail off into space until it is perhaps captured into the orbit of another planet. This is similar to atoms which have electrons clustered in orbit 'shells' around the nucleus. Applying an electrical force can change the orbits of these electrons, or knock them off the atom altogether. This movement of electrons is called an electric current.

In the following chapters, I'm going to lead you through the relevant science of water and its constituent parts of hydrogen and oxygen and then on to solid scientific principles that we can use to successfully achieve freedom from fossil fuels and filthy carbon pollution. What exactly are fossil fuels? What is carbon? What exactly is the HHO (oxyhydrogen gas) that people are adding to their engines to reduce consumption? What exactly is resonance and how can it split the water molecule? And lastly a look at the rather technical, but revealing patents and other documents that show us exactly how it all works in the real world. Read on and be a part of revealing this amazing technology to the world in a way that it will be impossible to conceal any longer or be suppressed by greed in any way.

1

The power of water

Water is all around us, even part of us, but what is it? and what power does it have? If we are going to discuss how to fracture water in order to use the power that is locked within it, then we must first understand what exactly water is. It's easy to believe that we know enough about water, but in fact, it still holds many mysteries, even for scientists.

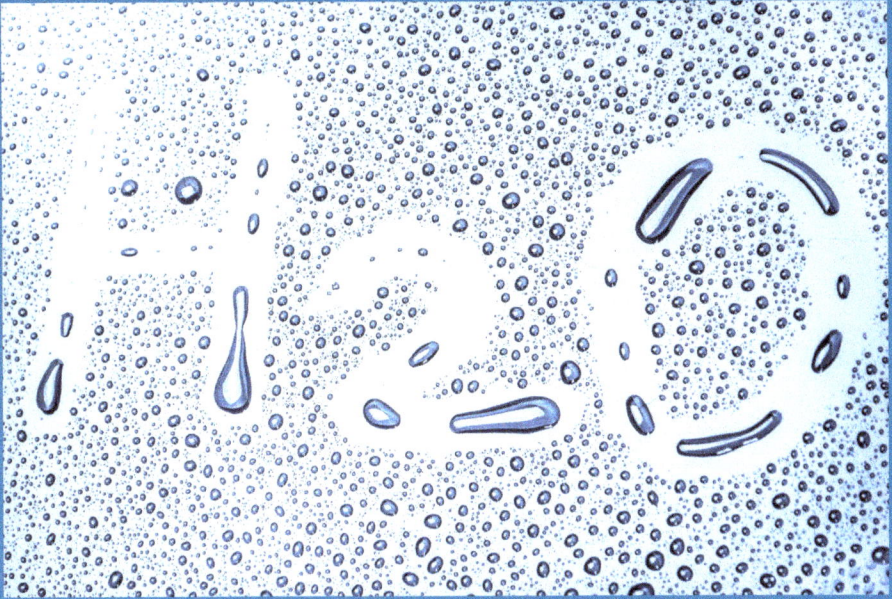

Water, the strangest substance on the planet

Water doesn't follow the usual laws of Chemistry! It has some unique properties. For example, it acts differently from most other liquids such as the way it floats in its solid-state (ice), in fact, it shouldn't really be a liquid at all on our planet at normal atmospheric temperature and pressure, as it is made of two very light gases, hydrogen and oxygen. It has a much higher heat capacity and low compressibility and can dissolve more things than any other common liquid. Water in its liquid form has a higher viscosity or 'surface tension' than other liquids due to the large number of *hydrogen bonds* that the molecule can form relative to its low mass. Since these bonds are very strong and difficult to break, it results in a much higher boiling point and melting point.

Why does hot water freeze faster than cold? Scientists simply don't know. Water molecules can flow *upwards* against the force of gravity because they stick to each other. That's called *capillary action* as in plants and the human body. Like trees drawing water up from their roots deep in the ground. It turns out that it's the second most common molecule in the Universe. At one time we thought that only the earth had water, but we now know that there is water on the Moon, Mars, Pluto, etc. It has been cycling, not only through the atmospheric system, glaciers, ice caps, rivers and oceans, but also through animals, plants, as well as humans for millions of years.

Scientists still struggle to understand it.

The properties of water

Water, precious, life-sustaining, powerful, sometimes dangerous and yet essential to life, is a fascinating substance with remarkable properties.

There are actually two distinct types of water, although you'd never know. They have an opposite atomic spin. Water is a very strange substance that seems to break the rules.

The formula, H_2O, means that water consists of two atoms of hydrogen and one atom of oxygen bound together. If we are going to use hydrogen, we must somehow separate the atoms. This is usually done by applying an electric current through the water. The process is called 'electrolysis', which is a simple experiment, performed in schools and college labs around the world. (See chapter 8 'Electrolysis'). However, as we shall see, *the traditional method is not efficient enough to be useful* and we must consider a different way of releasing the gases.

There may be readers who are concerned about using such a precious resource as water for energy on a large scale, as there are water shortages, particularly drinking water. But this technology can actually help with that, by reducing the cost of pumping or transporting water to where it's needed and also make the desalination of seawater cost-effective (over 97.5% of the world's water is saline). It can also help purify water.

Water is one of the most abundant compounds as about 71% of the Earth's surface is covered by water and water also exists in the atmosphere as water vapour, plus the rivers, lakes, reservoirs, polar ice caps, and glaciers, as well as in underground aquifers. It is also held within plants, animals, and human bodies! Not only that, but you would be making existing power stations no longer feasible and thereby saving the vast amount of water that they currently use. Another intriguing use of hydrogen power sourced from water is to recover the huge reserves of water found under the Sahara desert by NASA. It is currently too costly to recover with pumps using conventional fuels.

Where did the Earth's water come from?

It was thought until recently that all the water on earth was delivered by ice-bearing comets and asteroids, but a much more plausible explanation has recently been published in the 'Science' journal, that the earth is made up of hydrogen-bearing rocks, which when combined with oxygen through volcanic activity produced water. Therefore the right ingredients have been there since the formation of the earth[1].

Using water power

In about 3000 b.c., the Chinese used clocks incorporating dripping water to accurately mark time. In the year 1092, an inventor called Chang Ssu-Hsün created an astronomical clock in a tall bell tower, incorporating a water-wheel system to power the clock mechanism. It included animated characters ringing bells and banging drums. Water cannot be compressed and so can be used under extreme pressure as a cutting tool which will even cut, with a very fine precise point, through metal and other materials.

Look also at the power of water in coastal waves, ceaselessly eroding the rocks on our coastlines. The power of water is perhaps at its greatest in the dreadful tsunamis, caused by earthquakes, like the recent ones in Banda Aceh, Indonesia in 2004 and Japan, in 2011, resulting in a wall of water which is completely unstoppable, costing a great many lives and devastation on a massive scale. The Japanese tsunami also damaged several nuclear reactors, notably Fukushima, resulting in a major, ongoing, environmental disaster.

Hydroelectric

During the early industrial revolution, water was also used to power machinery. Water mills, driven by the weight of water cascading down a watercourse, were a very environmentally friendly way of providing power.

A steady source of water for mills was not always so straightforward, however, due to the intermittent rainfall and periods of drought. They were often supplemented with standing steam engines, or traction engines, connected to their line shafts by leather belts.

1. An unexpected source of Earth's water: https://www.science.org/doi/10.1126/science.aba1948

A modern-day extension of the water mill is the water turbine, sometimes on a massive scale in the great hydroelectric dams around the world. The principle uses a vast reservoir of water, created by damming up a steep-sided river valley, thereby creating a huge reservoir of water, which then drives the turbine as the water cascades down pulled by its weight under gravity. One novel power station Dinorwig, in North Wales, UK, is hidden inside a mountain, where water is released to drive turbines and produce electricity during peak demand and then pumped back up, usually during the night, using off-peak, cheaper electricity.

Steam Power

Early transport was powered by steam, the gaseous state of water. (Before that, of course, animals were used, particularly horses, and we still use the term 'horsepower' as a result). Steam power was used to power trains and a few early cars, as well as early cotton mills and farm machinery, although steam power is not so environmentally friendly as it requires the water to be heated up, usually by burning fossil fuels full of carbon, such as coal or gas.

Electricity generating plants (aka 'Power stations') even to this day are actually *steam-powered*. Believe it or not, we are probably still using electricity produced by steam-powered turbines in a coal-fired power station. (That is, of course, if not by alternative methods). Yes, your electric car may actually be powered by environmentally unfriendly coal and steam!

I have, in fact, worked in my capacity as a technical author, in several main coal-fired power stations. (Fiddlers ferry, in Cheshire, and Ferrybridge and Eggborough power stations, both in Yorkshire). The coal, (about 11 million tons per year, in just one power station!) fed constantly by long diesel-powered trains, is pulverised and blown directly into the furnace to heat the water. (One fact that often surprises people, is that during peak times of high demand, the power station sometimes has to use gas turbines to supplement the power output (read 'jet engines', yes, the same as used in aircraft).

Cooling Water

A huge amount of fresh river water or sometimes seawater is also required for cooling power stations, hence why they are usually built next to rivers. (About 12 million gallons per hour of wasted freshwater!). Why are they so thirsty? Because the steam coming out of the turbine needs to be cooled and condensed back to water and recycled through the huge pump house as shown in my illustration below. Hot water from the condensers is pumped into the eight 500-foot tall cooling towers to lose heat. Each one of these circulates 300,000 gallons of water per minute. About 8,000 gallons per minute are also lost to evaporation, seen in the steam escaping from the cooling towers.

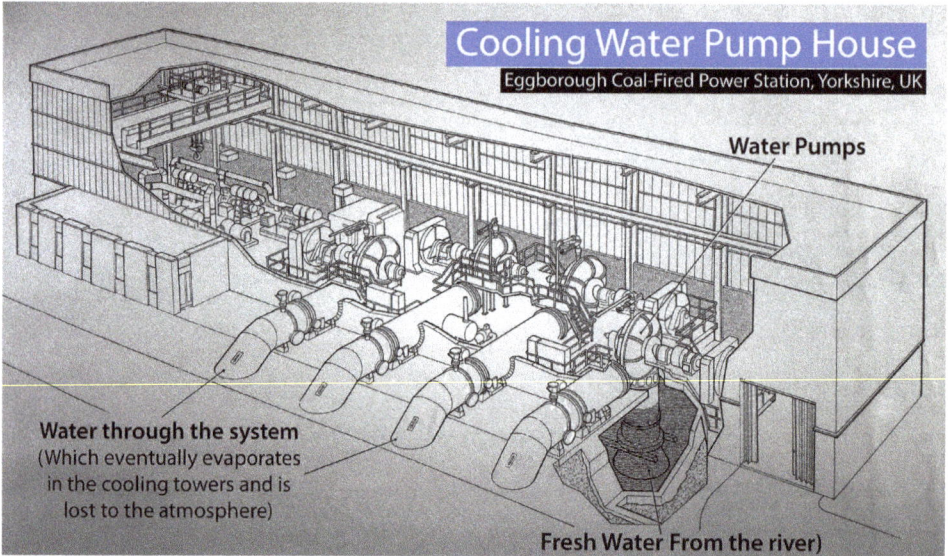

Cooling Water Pump House
Eggborough Coal-Fired Power Station, Yorkshire, UK

Water Pumps

Water through the system
(Which eventually evaporates in the cooling towers and is lost to the atmosphere)

Fresh Water From the river)

Cooling Water Pump house at Eggborough Power Station, Yorkshire, UK. Original in pen and ink by the Author.

One reason that this is a most inefficient way of producing power, is that it then has to be transmitted across many miles using overhead cables. (Hydrogen on-demand would totally eliminate this scourge of the landscape!).

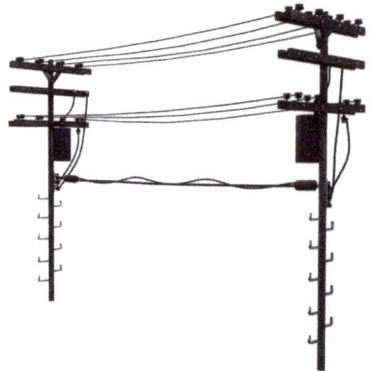

A lesson in efficiency from Stephenson's Rocket Locomotive

In 1829 the innovative 'Rocket' Steam locomotive by George and Robert Stephenson was a real game-changer. It won the £500 prize at the Rainhill Trials race and also the contract to run on the Liverpool and Manchester railway service. The race was designed to test the viability of the leading engines of the day. Hitherto steam engines had only been very large, heavy standing engines. People doubted that they could be made small and efficient enough to pull carriages along a railway line reliably. Instead of a single hot steam pipe within the boiler, it had twenty-five copper fire tubes which massively increased the effective surface area in contact with the hot water, resulting in greater efficiency. It also utilised a 'blast-pipe', which was a vertical steam jet directed up the chimney stack to help vent the exhaust smoke. This had the added effect of inducing a partial vacuum which pulled air through the firebox, drawing oxygen into the fire pit and massively aiding combustion. This forced draft, multi-tube design, became the basis of most steam engines for the next 150 years.

Rocket was designed by Robert Stephenson in 1829, UK.
Photo by Tony Hisgett from Birmingham, UK [1]

Even 'nuclear power' is really *steam-power*, as the electricity is in fact generated by huge turbines driven by steam produced by the cooling of the reactors.

1. Rocket Tyseley 3, Uploaded by oxyman, CC BY 2.0, https://commons.wikimedia.org/w/index.php?curid=15628107

This technology can help solve the water crisis.

Only 2.5% of the world's water is freshwater

Not only can you use polluted water or even seawater as fuel, but it can also dramatically reduce the cost of purifying and desalinating water!

Although water is one of the most abundant compounds, over 97.5% of the world's water is saline, covering about 71% of the Earth's surface as seawater. Freshwater is in short supply. Freshwater exists in the atmosphere as water vapour, plus the rivers, lakes, reservoirs, polar ice caps and glaciers, as well as underground.

Imagine the difference it will make, to be able to desalinate seawater and pump it cheaply to where it's needed without the high cost of fossil fuel involved.

The usually prohibitive, high cost of transporting water (which is very heavy) to where it's needed, will become almost zero!

Running pumps, trains, trucks, boats, tractors and other industrial or farm machinery can all benefit from almost zero-cost water!

Clean drinking water

Due to the rising populations and climate change issues, fresh water is becoming a dangerously scarce resource. But that's FRESH water. The planet is certainly not short of water, which is mostly brackish or saltwater. 326,000,000,000,000,000,000 gallons of it, (326 million, trillion gallons or 1,260,000,000,000,000,000,000 litres) covers most of our planet.

Although there are more than 300 million cubic miles of water on the earth's surface, being mostly saltwater, there are two billion people who do not have access to clean drinking water. However, hydrogen on-demand from water can address that problem in several ways. One, desalination is traditionally energy-intensive and therefore prohibitively expensive. Two, water is heavy and therefore expensive to transport using conventional means, usually requiring the use of fossil fuels. Three, pumping water from deep aquifers is also usually done with fossil fuel-driven pumps. All of these difficulties would benefit by using a little of the very water that they are processing, to run pumps and fuel the transport required.

Water purification

It takes ENERGY to convert salty seawater or polluted water to pure drinking water using desalination. That's where the technology of hydrogen on-demand can help. Not only that, but also the cost of pumping (or transporting in other ways) the water to where it's needed. Currently, apart from using solar or wind power to a certain extent, fossil fuels are used to provide this power. (Of course, solar doesn't work at night or during cloudy weather and wind power is also very variable and expensive to scale up and maintain). Water is of course what you are dealing with and therefore it makes sense to use the fuel capabilities it has at the same time.

Odd Fact

There is in fact a car that is designed to run on saltwater. It uses a technology called 'NanoFlowCell' by the Quantino Company. But they are keeping quiet about the details, which most probably uses some sort of nano PEM (Proton Exchange Membrane) fuel cell device. Using saltwater directly in an electrolysis process is otherwise dangerous as it produces chlorine gas.

H₂O
'Water molecule'
Rendered by the author in the open-source ray tracing app 'POVRay'

In short, if we source hydrogen from water efficiently:

- The usually prohibitive, high cost of transporting water (which is very heavy) to where it's needed will become almost zero!

- Pumps, trains, trucks, boats, tractors and other industrial or farm machinery can all benefit from almost zero-cost water!

- The 300 million cubic miles of water on the earth's surface, is potentially an inexhaustible supply of carbon-free fuel and yet, may also help preserve the freshwater that we need to survive.

Stephen Meyer, the surviving twin brother of Stanley Meyer (See Chapter 13) is also currently using his patented processes to purify water in Canada. The combination of Hydrogen and Oxygen (Hydroxy gas) has been shown to destroy organic pollutants. It is also highly effective against a series of pollutants including pesticides, pharmaceutical compounds, dyes, etc. It has been shown to be very effective in treating water and wastewater due to its high oxidation potential and nonselective nature. This is a far better way than introducing even more chemicals such as when using chlorine disinfection resulting in highly toxic by-products that can result after treatment.

Polluting groundwater by fracking (hydraulic fracturing)

There's currently a lot of controversy about fracking or drilling for fossil fuels by hydraulic fracturing, which now produces millions of barrels of oil each year. Fracking actually renders freshwater toxic, threatening the health of those living in the proximity of its operations, despite the fracking industry attempting to reassure everyone that it brings safe and clean energy. The method uses hazardous chemicals which are injected into very deep boreholes (more than a mile underground) under high pressure, causing toxic gases to be released from the rocks below. Polluting diesel-powered pumps used to inject the water, also release noxious chemicals and particulates into the environment. The chemicals used have been proved to be highly toxic and cancer-causing. Water used in the hydraulic fracturing process has leaked into drinking-water aquifers and other freshwater supplies. Also, the wastewater from fracking contaminates supplies.

The fourth state of water, 'EZ water'

Most people know that solid, liquid and gas, are the three states of matter, but there is another gel-like state of water, somewhere between a solid and a liquid which they are calling EZ water[1], meaning *Exclusion zone*. It forms in thin layers on surfaces or other boundaries of water, such as where the water meets the air or *other materials.*

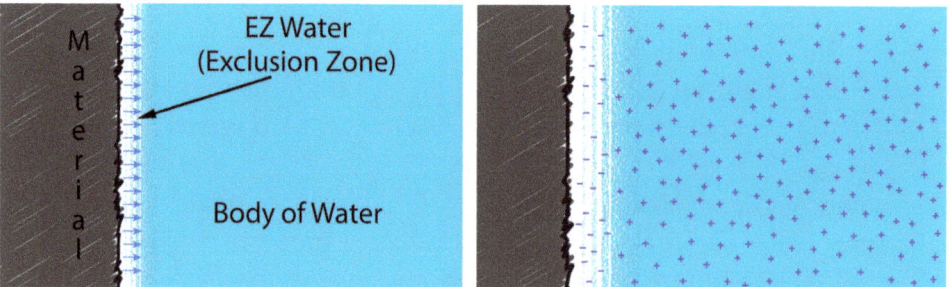

The thin 'exclusion zone' pushes everything out and becomes negatively charged.

1. TEDx University of Washington Dr Gerald Pollack:
    https://www.youtube.com/watch?v=i-T7tCMUDXU

The 'exclusion zone' tends to push away any impurities, even pollution and bacteria. In seawater it even pushes out the salt, leaving a layer of fresh water. Although this layer is only the thickness of a human hair, it can be built up in multiple layers and so can actually be utilised for desalination. It goes a long way to explaining 'surface tension' and why water tends to stick together in various ways. If you place a paper clip on the surface it will float and yet just beneath, it sinks. Water vapour in clouds will come together as raindrops and there's even a lizard that can run across water called a 'Jesus lizard'.

All this is possible because negative charges keep on building up along the boundary layer and it has been found that it's the energy from light, particularly infra-red, which is feeding it with this charge and expanding it, in the same way that photosynthesis feeds plants with photons.

Negative charges building up around water droplets within a field of positive charge means that these objects, although similar charged, which we would normally assume to repel each other, are actually attracted (stick together) because of the buildup of excess positive charge between them. This electric potential has been shown to be useful as 'free energy' which can even be harnessed.

As we might expect with a substance that has a different molecular structure, the properties are also different. Indeed EZ water has a molecular structure of a hexagonal lattice, with six hydrogen atoms connected with oxygen atoms in between. It is in fact not H_2O! It's H_3O_2.

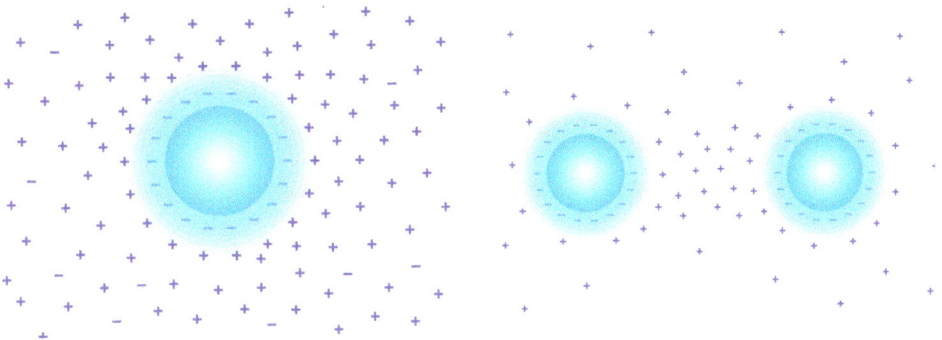

Particles in the water will also be surrounded by an 'exclusion zone' of negative charge. If two such particles are in proximity they will be attracted due to the build-up of a shared positive charge between them.

Why this matters to the idea of splitting water is that all the action of gas formation happens within that very boundary layer.

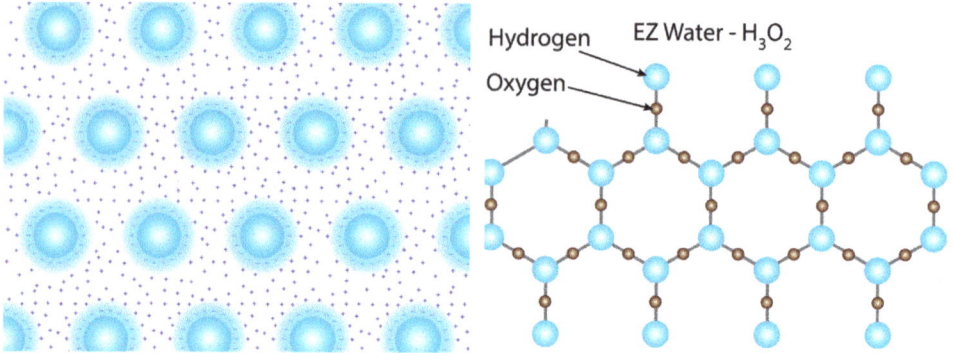

Hydrogen   EZ Water - H_3O_2

Oxygen

A lattice of EZ water. Due to the hexagonal matrix, it is not H_2O but H_3O_2

Fire underwater

Extreme Plasma voltages can help to create water plasma. Perhaps you learned at school that everything has three states of matter, solid, liquid and gas? Well, plasma is the fourth state. It's something like an energised gas. Water in that state is easier to fracture and since it is already ionised it helps make the process more efficient. Another principle of magnetism and electric induction that we will use to split the water molecule, is that of alternating currents and electric fields. In other words, that way we can rapidly switch the direction of the magnets or currents to impose a force on anything caught within their respective fields. It can have the effect of flexing the water molecules back and forth until they break. (More on Plasma electrolysis in Chapter 16).

So you see. Maybe we don't know so much about water after all. In the next two chapters, I'll look at its component parts oxygen and hydrogen separately, to see what it's really made of. In effect a highly explosive gas and an oxidant that will massively aid combustion.

2

Oxygen

As we are talking about splitting water into hydrogen and oxygen, let's take a closer look at them in turn.

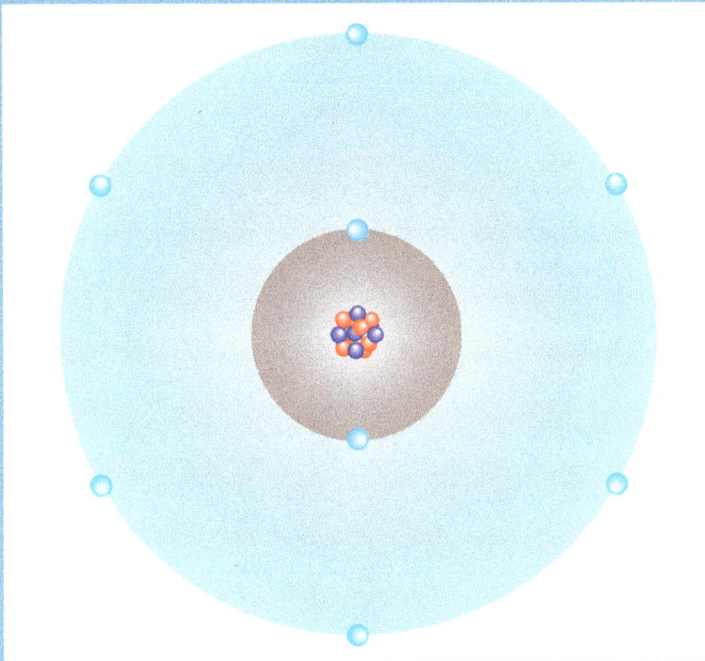

Oxygen is in the air we breathe, but only about 21% as the rest is mostly nitrogen. This has consequences for engines as the more oxygen we can get in, the better the combustion. If we can use water as a source of fuel, guess what? It's already one-third oxygen, which means that it is the perfect combination for efficient combustion and an additional bonus is that its carbon-free!

Oxygen is in fact the most abundant element on our planet, in combination with other elements in the air that we breathe, in water and in organic compounds within all forms of life, including our own. After hydrogen and helium, it is the third-most abundant element in the universe. Oxygen atoms easily combine with carbon atoms to make carbon dioxide, which plants breathe. The oxygen gas molecule currently constitutes about 21% of the Earth's atmosphere, though this figure is highly variable and changes over periods of time. Oxygen also makes up about half of the Earth's crust in the form of oxides.

The name 'Oxygen' comes from a combination of the Greek words 'oxys' meaning "acid" and 'genes' meaning "forming".

One-third of water is made of Oxygen. It is colourless, odourless and tasteless. That's the 'O' in 'H_2O' although normal atmospheric oxygen is a diatomic gas (O_2), meaning that the molecule is made of two atoms of oxygen. The oxygen atom has eight neutrons and eight protons and is surrounded by eight orbiting electrons, which makes it **heavier than a hydrogen or a helium atom, and even heavier than a carbon atom.** We know it usually as a gas, but it can also be usefully compressed into liquid form in order to transport it easily and use it in industrial processes, such as cutting and welding and medical uses in hospitals. Liquid oxygen combined with hydrogen is also used as a rocket propellant.

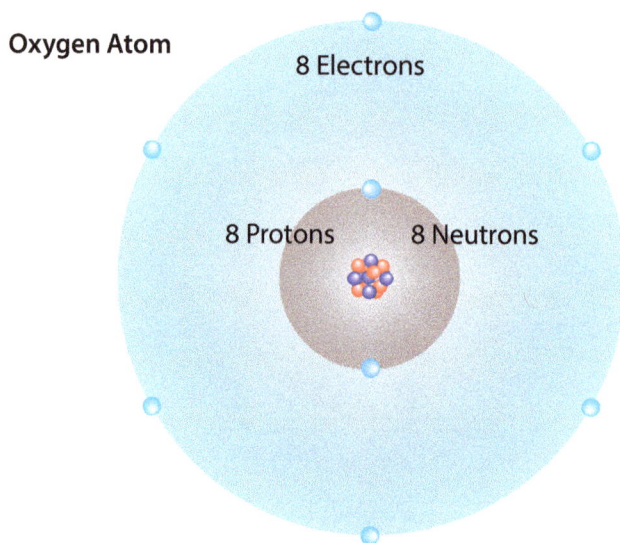

Oxygen Atom

8 Electrons

8 Protons   8 Neutrons

A diagrammatic representation of an oxygen atom, by the author.

Oxygen Molecule

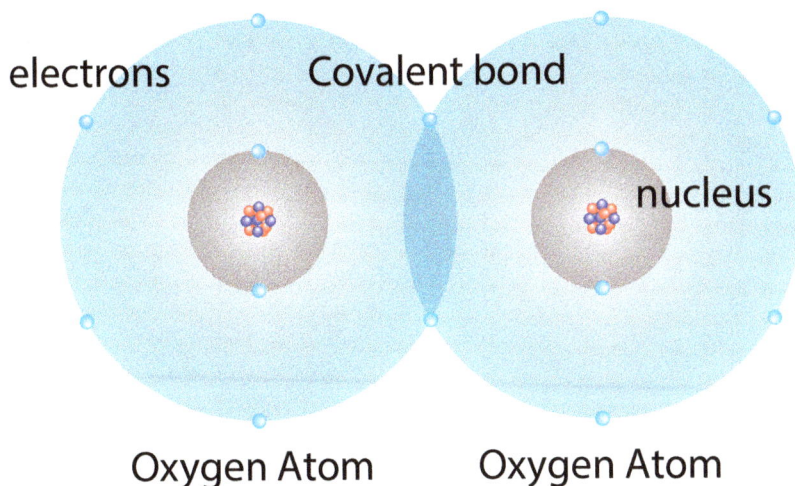

*Two oxygen atoms combine to create the oxygen molecule (O_2).
At standard temperature and pressure, two atoms of the element bind
together to form the dioxygen molecule, a colourless and odourless diatomic
gas. Illustration by the author.*

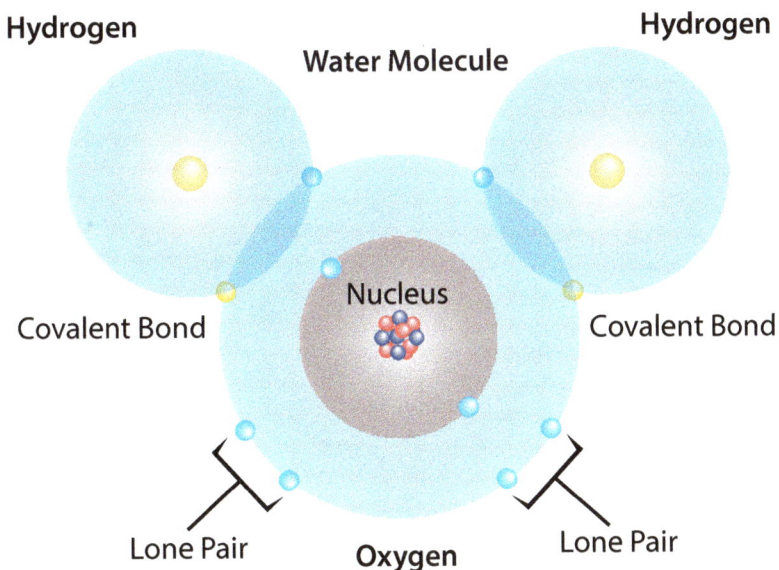

*How Hydrogen combines with oxygen in water, sharing 'covalent' bonds.
Illustration by the author*

We can, of course, breathe pure oxygen, and bottled oxygen can be useful in situations where the air is thin at high altitudes. And for example, when climbing mountains or ascending in aircraft and also in lifesaving medical situations when breathing is impaired. Oxygen in the earth's atmosphere is unstable because it can easily form compounds with almost all other elements. (For example combining with carbon to create carbon dioxide (CO_2), or carbon monoxide (CO), so it needs to be constantly replenished by the planet's greenery).

Creating oxygen

Photosynthesis is the process by which plants use sunlight, water, and carbon dioxide to *create oxygen* and energy in the form of sugar. (Sugar being a **carbo**hydrate is also composed of carbon, oxygen and hydrogen and can be explosive. It can even be used as a rocket propellant!).

Plants require three things to perform photosynthesis, carbon dioxide, water, and sunlight. Carbon dioxide enters through tiny holes in all parts of the plant, its leaves, flowers, branches, as well as stems and roots. Plants also, of course, need water to make their food, mainly provided by rain. (Note that the rain is also produced by the energy from the sun. As water evaporates and

Rainforests are the source of a great deal of the world's oxygen

becomes water vapour, or 'clouds', the uneven air temperatures cause pressure differences and air movements. These winds blow them around and raindrops form from the water vapour as the droplets grow in size by combining together, particularly around dust or particles of pollution).

The green colour of leaves is made by a pigment called Chlorophyll and is essential to the photosynthetic process. At the end of the season, the plant's leaves turn off the production of this pigment, returning back to their natural brown colour, before dying off.

By using the power of the solar radiation, the plant breaks down any available CO_2, extracting the Carbon and releasing the Oxygen, enhancing the atmosphere in the process. This way, whole forests are locking in that energy as a form of carbon, ultimately becoming oil, coal, gas and methane, which is given off when vegetation rots.

Oxidation

The process by which oxygen unites with other substances is called 'oxidation' When that happens it forms an 'oxide'. For example, when oxygen combines with iron and water to create iron oxide, better known as 'rust'.

Rust, or 'iron oxide' claims yet another victim.
Saltwater accelerates the rusting process.

Liquid oxygen

Liquid oxygen, or LOX, is the liquid form of molecular oxygen and is widely used in the aerospace, submarine and gas industries.

It's also used as the oxidiser in liquid-fuelled rockets, along with liquid hydrogen, such as in the Space Shuttle main engines.

Liquid oxygen with liquid hydrogen used in the Space Shuttle's main engines.

Solid oxygen

On the other end of the temperature scale, at very low temperatures of about −183 °C (−297 °F), oxygen becomes a solid. This can also be very useful, especially as it has magnetic properties.

A good supply of oxygen

As any petrolhead will tell you, with internal combustion engines, the more oxygen you can get to pass through, the more power you will get out of the engine. Air filters generally restrict the flow of air, so special racing versions can be fitted to help 'aspirate' the engine. The problem is that the atmosphere is only about 21% oxygen. However, if you are using water as a source of hydrogen to provide combustion, then oxygen is a major component as it is. It helps the highly explosive hydrogen to burn more furiously.

So the secret to good combustion is a good supply of oxygen. My grandmother taught me that. Yes, she showed me how to build a coal fire and place a newspaper over the whole fireplace opening to 'draw the flames through' until the coal started burning properly. (Coal, which is mainly carbon, is very difficult to light!). There was also a special vent that you could open which let more air through the grate until the fire was burning properly. Conversely, to put out a fire, one effective technique is to 'smother' the flames. This has the effect of starving the flames of oxygen, after which combustion is impossible. A 'fire blanket' is sometimes provided for just such a purpose. When using fire to smelt metal, very hot temperatures are required and bellows are sometimes used to get more oxygen into the flames to bring up the temperature.

Nitrous oxide, no laughing matter.

Now, under pressure, a nitrogen atom can attach itself to four oxygen atoms creating a nitrous oxide molecule. The nitrogen then becomes the carrier of oxygen. This gas, more commonly known as laughing gas, when breathed in raises the pitch of a person's voice and causes giggling fits. It has serious medical uses though, especially for surgery and dentistry, due to its anaesthetic and pain-reducing properties. But the reason I'm mentioning it here is its ability to deliver oxygen directly to an engine's intake.

One of the most highly tuned forms of engines is those used by drag-racers or hot-rods and one of the secrets to their incredible performance (speeds of 0-100mph in less than one second), is that the engine, instead of being 'naturally aspirated' with ambient air, which only contains about 21% oxygen, is injected with nitrous oxide. If you can imagine a room full of air being compressed down to the size of a football, it would then be in liquid form at high pressure. This could be contained in a tank and released through a valve which would then allow it to expand again rapidly. If this gas was nitrogen and oxygen, instead of air, it would compress to its liquid form nitrous oxide.

When this is released into an engine's air intake, it delivers almost pure oxygen (NO_4), which, as any car racing enthusiast will tell you, massively improves combustion and therefore performance. Now, since fractured water releases a combination of hydrogen and oxygen, in the proportion of 2:1, that surely will have a similar effect. You simply don't need to add extra oxygen as the fuel released already contains it, in perfect proportions. The relatively high proportion of oxygen, compared to the low amount in ambient air, will create the perfect mix for efficient combustion. It leaves no unburned gases, no dangerous cancer-causing particulates, no engine-clogging carbon deposits, just a pure hydrogen flame, with plenty of oxygen delivered along with it for highly efficient engine performance. Couple that with an engine optimised to run on hydrogen and you can forget all other alternative fuels with their associated problems. Also, no 'range anxiety' as you get with electric cars.

As water has three times the energy of fossil fuels, a smaller amount would be needed with less weight. Plus, no dangerous storage of combustible gases as the water can be stored safely and only turned to fuel as and when it's required. Expensive batteries, using up the earth's precious metals, and complex, expensive fuel cells will certainly seem non-viable in comparison.

A flame is an 'exothermic' chemical reaction which gives off energy in the form of heat and light. Here, ignited by the heat of friction, the phosphorus on a wooden match tip is reacting with the oxygen in the air.

So why doesn't water burn?

It may seem like an odd question, but the reason is that it has a high boiling point (at 100 °C/ 212 °F), melting point and viscosity. (Scientifically speaking, that is because of the high number of hydrogen bonds that each molecule can form, relative to its low molecular mass). In other words, it turns to steam, or water vapour, before it gets hot enough to ignite, at its 'flashpoint'. That is where electrolysis comes in. Instead of heating the water to burn it, we can run electricity through it to break the molecular bonds.

Burning involves the oxidation of a combustible substance. When a combustible substance burns, a chemical reaction occurs in which the substance combines with oxygen, usually provided by the air and gives off heat, gases, and often light in the form of flames. However, if you use a plasma spark and do it quickly enough, then yes, it will burn instantly (See chapter 16 on Plasma).

Adding water can actually make other materials more volatile, because of its hydrogen and oxygen content. The massive explosion on the ship in the Lebanese Dock on 4 August 2020, was due in part to the large amount of fertiliser (ammonium nitrate) stored at the Port being damp and therefore highly unstable[1].

The combination of hydrogen and oxygen gas, oxyhydrogen, has some special qualities and that's because it's the perfect combination for combustion in itself. Yes, it has everything it needs for standalone burning. For example, it will burn perfectly well underwater. However, the water doesn't heat up! This is proof that it doesn't need any elements from its surrounding environment. It doesn't react with water or the air when out of water for that matter. The flame only reacts when it comes into contact with a solid object, such as a stone, which it can even melt underwater. Even metals will react underwater with an oxyhydrogen flame- steel, tungsten, gold, etc.

This means that it can be used for cutting and welding underwater. When the oxyhydrogen flame is out of the water it reacts with other materials by recombining back into water. Burning lime, for example, raises the temperature up to nearly 5000 °F (2760 °C) producing a huge amount of heat and light in the process. (That's where the word 'limelight' came from, as it was used in

1. An Investigation into the August 4 Beirut Blast:
https://en.wikipedia.org/wiki/2020_Beirut_explosion

theatre floodlights at one time). When it does not react to other materials, the oxyhydrogen flame is only 140 °F (60 °C) about a tenth of the heat of a candle flame.

All this is even more proof of the tremendous amount of energy that is within the water and the resulting oxyhydrogen gas that is released when it is fractured.

Fire, ice and water

Combustion is a chemical reaction when oxygen atoms react with hydrocarbon molecules or other elements, and which usually gives off a great deal of heat and light in the form of flame. It is in fact very rapid oxidation, otherwise known as 'burning'. When oxygen reacts with hydrogen gas during combustion it results in water molecules. (H_2O).

If you mix pure oxygen with a fuel gas (most commonly acetylene, a hydrocarbon), it produces a high-heat, high-temperature flame that can be used for welding. The oxy-acetylene torch is powered by the gas which has been compressed into pressurised steel cylinders.

Welding with an oxy-acetylene torch, also known as 'gas' welding.

Atmosphere

The thin blue line

Our atmosphere (from the Ancient Greek words 'atmós' (vapour, steam), and 'sphaîra' (sphere), sometimes called the 'thin blue line', is mostly made of nitrogen (78%) and only about 21% oxygen. The other gases are argon (0.9%), carbon dioxide (0.04%) and other trace gases of neon, helium, methane, krypton, ozone and hydrogen, as well as water vapour.

Although it appears 'thin' when looking at images of the earth, according to NASA, the upper atmosphere (called the 'exosphere') actually extends up

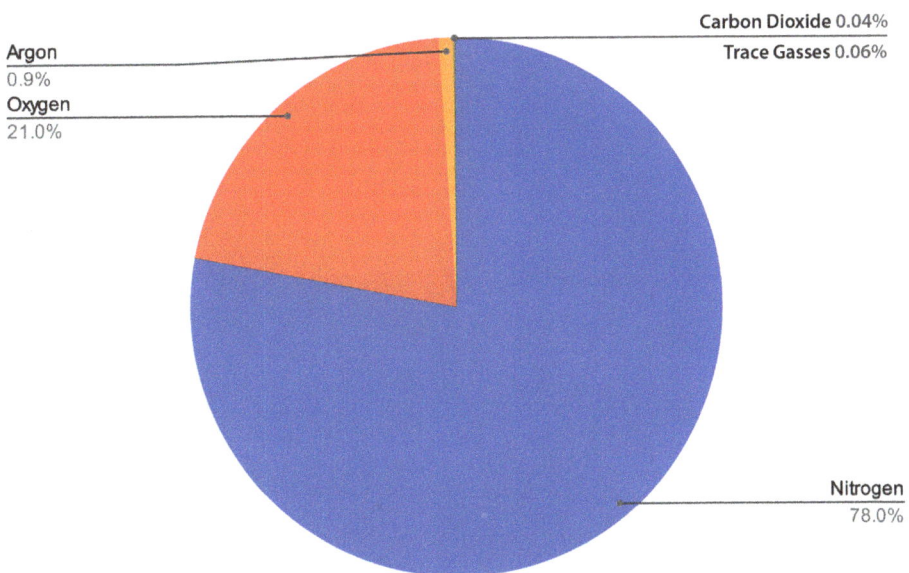

Carbon Dioxide 0.04%

Trace Gasses 0.06%

Argon
0.9%

Oxygen
21.0%

Nitrogen
78.0%

to about 6,200 miles (10,000 km). However, what is normally regarded as the limits to our atmosphere/beginning of space, is about 62 miles (100 km).

The air pressure decreases with altitude, and at sea level, is about 15 pounds per square inch (1 kilogram per square centimetre). Above 10,000 feet (3 km), the air is so thin that anyone flying that high needs to carry extra oxygen in a tank to help them breathe.

When animals and humans breathe, they take in oxygen from the air and give out carbon dioxide (CO_2), which plants can use to make food. That process is called 'photosynthesis'. In return, plants "suck in" the CO_2, retain the "C" for growth, and give off a large amount of O_2.

So where does hydrogen figure in all this? Hydrogen is barely present in our atmosphere, which contains only about 0.00005%. However, it is combined within many other elements on our planet, such as water and hydrocarbons. Consequently, if we need to use it, we must extract it from any of those elements. Let's look at Hydrogen next.

3

Hydrogen

The molecule H_2 is 'diatomic', meaning that it consists of 2 atoms of hydrogen. (The electron of one is attracted to the nucleus of the other, locking the two together). It is the lightest and most abundant element in the Universe, colourless, odourless, non-toxic but highly flammable.

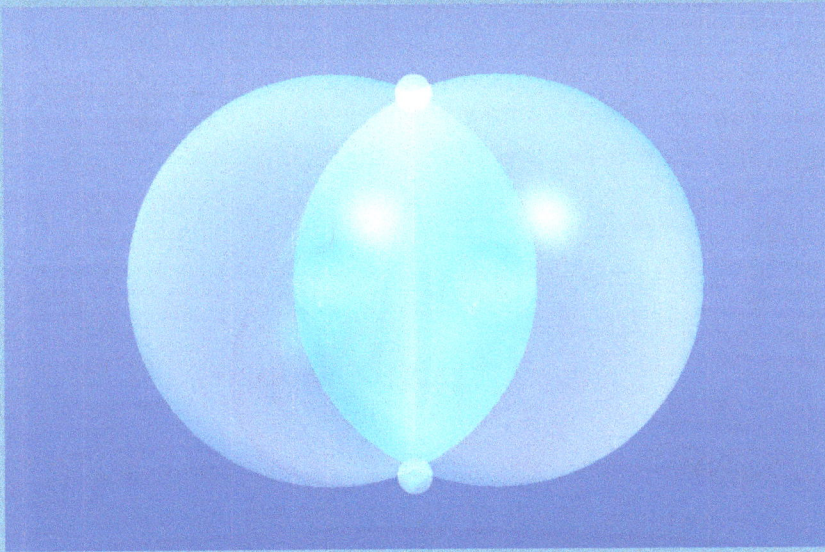

We know that hydrogen is highly explosive, especially after disasters such as the Hindenburg airship and the massive explosive potential of the H bomb. It has also been regularly used, in combination with oxygen as rocket fuel by NASA to power the massive booster rockets of the space shuttle.

Hydrogen - tiny, yet abundant.

Since we are talking about splitting water to release hydrogen, to use as a fuel, perhaps we should take a closer look at the properties of hydrogen?

Hydrogen, the most abundant element in the Universe and also the simplest, is a colourless, odourless gas and is easily ignited. Once ignited hydrogen burns with a pale blue, almost invisible flame, that is difficult to see in daylight. It burns with a very clean extremely pure flame and can be used for cutting and welding metal. (It is also used as rocket fuel, but more about that later). As a lighter-than-air gas, it has been used in the past to fill man-carrying balloons and airships, such as the infamous Hindenburg and R101, both of which ended in tragedy when they exploded.

It is the smallest molecule, (so small in fact, that it can pass through steel containers, which proves that storing hydrogen is to be avoided!) and has only one electron surrounding the nucleus of each of its two atoms that make up the molecular form of hydrogen. Our atmosphere contains only about 0.000055% hydrogen, and we are in fact breathing a little bit in, so it is obviously not toxic.

Hydrogen occurs naturally on earth but only combined with other elements in liquids, gases, or solids. Hydrogen combined with oxygen is water

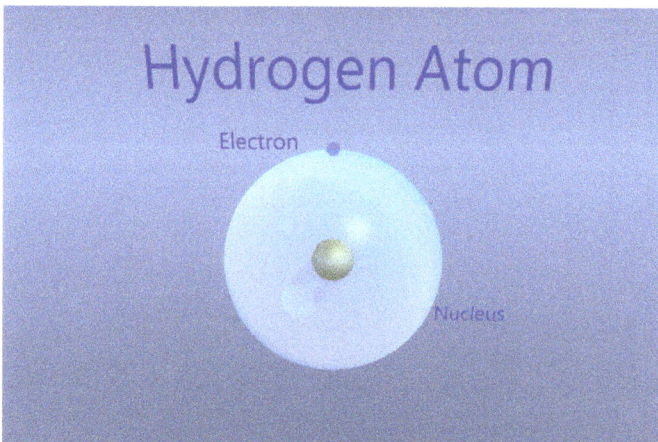

H 'Atom' Rendered by the author
in the open-source raytracing app 'POVRay'

Hydrogen, H2, (composed of two atoms) is the lightest and most abundant element in the universe. It is colourless, odourless, but highly flammable.

(H$_2$O). Hydrogen combined with *carbon* forms different compounds called '*hydro*carbons' found in fossil fuels such as natural gas, coal, and petroleum.

It is rare to find natural hydrogen on earth as it's always in combination with something else, but it can be obtained in other ways. It can be extracted from fossil fuels, such as methane gas, but also from water, using a process called electrolysis as is the theme of this book. Pure hydrogen already has many applications as a clean zero-carbon fuel. It is already being trialed in a range of different types of vehicles to directly replace fossil-fuels, or create electricity in fuel-cell vehicles.

Our hydrogen-powered sun

Stars such as the sun consist mostly of hydrogen. The sun is essentially a giant ball of hydrogen and helium gases. So how does it burn in the vacuum of space without oxygen? The answer is that it burns through a different process called 'nuclear fusion' where the atoms are fused together to become 'helium' due to the immense pressure caused by the giant star's massive gravity.

The sun is the true source of most of the earth's energy such as that in fossil fuels due to photosynthesis when the ancient forests were growing. The dying forest's organic matter locks energy into the layers of earth as carbon deposits.

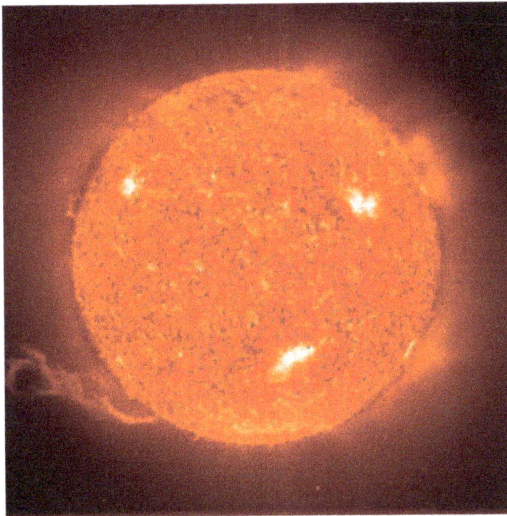

The sun is essentially a giant ball of hydrogen gas undergoing fusion into helium gas. This process causes the sun to produce vast amounts of energy which is the true source of most of the earth's energy.

Obtaining hydrogen

The way we produce hydrogen is classified into colour-coded categories, green, blue, grey, even pink and yellow, depending on how 'environmentally friendly' each process is. As follows...

Green hydrogen (The cleanest)

Green hydrogen can be produced by splitting water with a process called electrolysis. This produces only hydrogen and oxygen. We can use the hydrogen and vent the oxygen to the atmosphere with no negative environmental impact.

However, electricity is required for the process of electrolysis, but green hydrogen can be powered by renewable energy sources, such as wind or solar, making green hydrogen a clean solution.

Grey hydrogen: (The cheapest)

This is the cheapest way to produce hydrogen, so most of the hydrogen that we use (around 95%) is produced this way. It involves splitting methane, which is composed of carbon and hydrogen, using a process called 'steam methane reforming' (SMR).

However, this process releases carbon emissions.

The carbon from Grey hydrogen can be captured and buried, in which case it is then called blue hydrogen. (Confusing, isn't it?).

Blue hydrogen (Also makes carbon dioxide).

Blue hydrogen is produced by splitting natural (i.e. fossil-fuel) gas into hydrogen and Carbon Dioxide (CO_2) either by Steam Methane Reforming (SMR) or Auto Thermal Reforming (ATR), but the CO_2 is also useful, so captured and stored. Since the resulting Carbon Dioxide is captured, it lessens the possible environmental damage.

The H bomb.

While we are talking about energy and hydrogen in particular, I cannot omit the truly awful devastating power of the hydrogen bomb. Many thousands of times more powerful than the atom bomb, thankfully the only times it has been detonated is in tests. The largest ever, the Cold War Russian Tsar bomb, the most powerful nuclear weapon ever created, unleashed a massive amount of destructive power equal to about 58 megatons of TNT which, when tested, on 30th October 1961, sent a mushroom cloud up to 213,000 feet, or 65 km high into the atmosphere.

The way this relatively small device works is by utilising radioactive uranium 235 to cause nuclear fusion between hydrogen atoms, crushing them together to make helium, the same way as it happens within the sun.

The temperature within a hydrogen bomb explosion is hotter than the sun, so I have just one question, when did someone 'put that much energy in, to get that massive amount out'? How does that work with the 'Law of thermodynamics'?

Explosion of a hydrogen bomb. (Einstein, through his formula, $E=MC^2$ showed that the energy hidden inside the hydrogen atom is enormous)

The Hindenburg disaster

The explosion of the Hindenburg as it came into land in Manchester Township, Ocean County, New Jersey, the USA on 6th May 1937, sounded the death knell on the golden age of airships. The Hindenburg was at the time the world's largest aircraft and was filled with hydrogen as it is lighter than air and provided its buoyancy. Unfortunately, hydrogen is also highly flammable and attempting to land in a thunderstorm was provided with a spark of lightning and a rain-soaked rope which provided a perfect path to earth for the built-up electrical charge.

One controversial theory is that the explosion was aided by the silver doping that waterproofed the skin and silver paint designed to reflect the sun to keep the envelope cool, although I think that the Jury is still out on that one.

Hydrogen in summary

Hydrogen is the simplest element. Each atom of hydrogen has only one proton. Hydrogen is also the most abundant element in the universe.

Stars such as the sun consist mostly of hydrogen. The sun is essentially a giant ball of hydrogen and helium gases.

Hydrogen occurs naturally on earth only in compound form with other elements in liquids, gases, or solids. Hydrogen combined with oxygen is water (H_2O). Hydrogen combined with carbon forms different compounds—or hydrocarbons—found in natural gas, coal, and petroleum.

- It is represented by the symbol H.

- Is lighter than air and has the highest energy content for fuel by weight.

- Is extremely flammable, needing just a 4% concentration in the air!

- It makes water once the hydrogen shares a covalent bond with oxygen.

- Stars are mostly made of hydrogen. An example would be our own sun.

- It has many applications and great potential as a clean fuel as it can be used for hydrogen-fuelled vehicles.

- It is rare to find natural hydrogen on earth but it can be obtained in other ways including electrolysis from water, as is the theme of this book.

Hydrogen has such a high chemical reactivity that the thermal impulse (e.g. a spark) needed to ignite it is about 10 times less than that necessary to ignite an equivalent mixture of air and natural gas. A concentration mixture of anything more than 5% hydrogen in air usually ignites with an explosion.

Hydrogen is an excellent energy carrier with respect to weight. 1 kg of hydrogen contains 33.33 kWh of usable energy, whereas petrol and diesel only hold about 12 kWh/kg (see www.h2data.de).

So hydrogen is a super fuel, especially in combination with oxygen.

Next, we'll look at the ins and outs of running an engine on hydrogen. Oh, by the way, most engines already do, except in combination with carbon.

4

Running an engine on hydrogen

Did you know that your engine is already running on hydrogen?

Whether an engine is fuelled by petrol, diesel, LPG gas, heavy oil (eg. Marine) Avgas (aviation fuel), or biofuel,

they are all running primarily on the hydrogen within it.

The Carbon content really doesn't help!

ICU engines are not very energy efficient. When burning hydrocarbon fuels they only convert about 25% to mechanical energy, whereas hydrogen fuel can more than double that.

Saving the Internal Combustion Engine

Believe it or not, the very first internal combustion engine (ICE) ran on hydrogen and oxygen. The Franco-Swiss inventor Isaac de Rivaz, in 1804 after designing several steam engines, successfully used an explosive charge of hydrogen and oxygen within the cylinders of a stationary piston engine. By 1807 he successfully used it to propel a carriage some distance thereby inventing the world's first internal combustion-engined vehicle.

Hydrogen has been successfully used as a fuel in all manner of vehicles from cars to buses but not without its problems. Firstly, hydrogen is notoriously difficult to store and transport safely. Being the smallest molecule, it is even capable of seeping through stainless steel. Storing it safely and securely under pressure as a liquid, similar to the way LPG (liquid petroleum gas) is managed, in order to more easily dispense and distribute it, brings further difficulties. The infrastructure isn't there either, especially as hydrogen cars are relatively expensive and therefore not yet ready for the mass market. That could change if hydrogen can be sourced efficiently from water, negating the need for refuelling stations or on-board storage, especially if one of the breeds of smaller, cheaper new engines, developed just for hydrogen are also available.

Most engines already run on hydrogen.

What we may not realise is that every engine that runs on petrol (called 'gas' in the USA), diesel, LPG (Liquid petroleum gas), biodiesel, and even avgas (aviation fuel), is actually running on hydrogen. Hydrogen bonded to strings of filthy black carbon to be precise. Hydrogen that burns with a pure hot flame and carbon, which is difficult to light, combusts poorly and clogs all the delicate moving parts with unburned particulates. So we've been using hydrogen all along, but spoiling it with the carbon that it's bonded to. Hydrocarbon fuel is just that, hydrogen and carbon. Water is similarly composed of hydrogen which does the work but also oxygen which acts as an oxidant and aids combustion. As any petrolhead will tell you, improve the aspiration of the engine and you will increase performance. Which is similar, in a way, to using bellows on a fire. Flames need oxygen. Since there is only

21% oxygen in the air, hydrogen delivered with a good amount of oxygen by splitting water into its constituent parts of 2 molecules of hydrogen to every molecule of oxygen is a win-win situation with the only by-product being fresh water! However, if you want to run an engine on hydrogen alone, you need to pay attention to some important differences. As hydrogen burns more efficiently and importantly, more quickly, the engine must be tuned to respond to those conditions.

Don't engines also need oil for lubrication?

Advances In nanotechnology and ceramic coatings to reduce friction means that engines may not even need oil. It also leads to better efficiency as heating and wear are controlled. These processes are already being introduced to formula 1 and other advanced engine fields.

Believe it or not...

The world's first internal combustion engine used hydrogen gas for fuel and possessed some features of our modern engines such as spark ignition.

In fact, in 1970, inventor Paul Dieges was the first to patent a modification to gas-powered internal combustion engines, which enabled them to run on hydrogen.

A = Cylinder,
B = Spark ignition
C = Piston
D = Balloon of hydrogen fuel
E = Ratchet
F = Opposed piston with air in and exhaust out valves
G = Handle for working opposed piston.

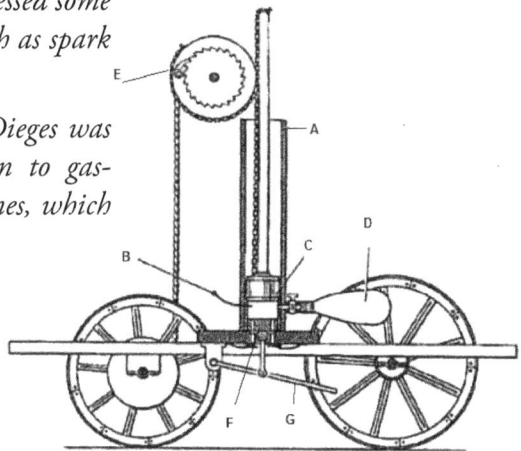

François Isaac de Rivaz 1752 – 1828) 1804 patent in the Republic of Valais

Implosion, not explosion.

Normally the air and vaporised fuel mix will explode, but with hydrogen, it actually implodes as it returns to water. The most efficient hydrogen-fuelled engines need to take that into account also.

Ignition timing

One such consideration is ignition timing. The correct ignition timing is crucial to the performance of an engine. That is the timing of the spark that ignites the fuel mix, causing the explosion, which pushes the piston down to rotate the engine. This is called 'advancing the ignition'. Advance is required because the air/fuel mixture does not burn instantly. It takes time for the flame to ignite the mixture.

'Top dead centre' is the point when the piston of the number one cylinder in an engine is at its highest point, and on the compression stroke of a typical engine's four-stroke cycle. When the piston is at TDC (top dead centre) position it needs to rotate a few degrees more before the fuel/air explosion pushes it down again. There's a slight lag in the combustion of fossil fuels, requiring a timing adjustment on the engine of about 10 degrees. However, with hydrogen combustion, the ignition is instant and if the ignition is advanced then the piston may be pushed backwards, damaging the engine.

In an internal combustion engine, using fossil fuels, the combustion cycle is very fast, only about 0.007 seconds. Most of the fuel molecules are too large to burn completely in this extremely limited time. The situation is made worse by the fact that the spark plug only ignites a small percentage of the fuel. The explosive force generated must cascade from one fuel molecule to the next as it propagates through the combustion chamber of the engine, wasting precious time.

Engines designed especially for hydrogen.

The best way is perhaps to develop new engines, especially for hydrogen, and during my research I have found two brand new engines which have been optimised to run on hydrogen.

Aquarius Engines

What Aquarius Engines in Israel (https://www.aquariusengines.com/) have produced is incredible. The new, lightweight engine, the Aquarius Two-Sided Free Piston Linear Engine (FPLE), weighing just 10kg (about 22 pounds), was specially developed to run on hydrogen. The highly efficient engine has a very streamlined design, is very compact and has only one moving part, its piston. It's so inexpensive to build and maintain compared with traditional engines as it consists of only about 20 parts. It was primarily designed to generate electricity for plug-in series hybrids or range-extenders. The engine drives a generator to charge a battery or run a traction motor in a hybrid electric vehicle.

The engine is air-cooled and needs no lubrication but instead has graphite piston rings and special coatings to reduce friction. Because there is **no oil used**, there is no chance of leaks or for that matter, unburned hydrocarbons being produced during combustion as well as simplifying servicing. It drives a generator that feeds the battery or traction motor of a series hybrid with electricity. It could even be used as an auxiliary power supply (APU) in large trucks and marine applications, for use in drones or as auxiliary power supplies in aircraft.

Free Piston Linear Engine-Generator

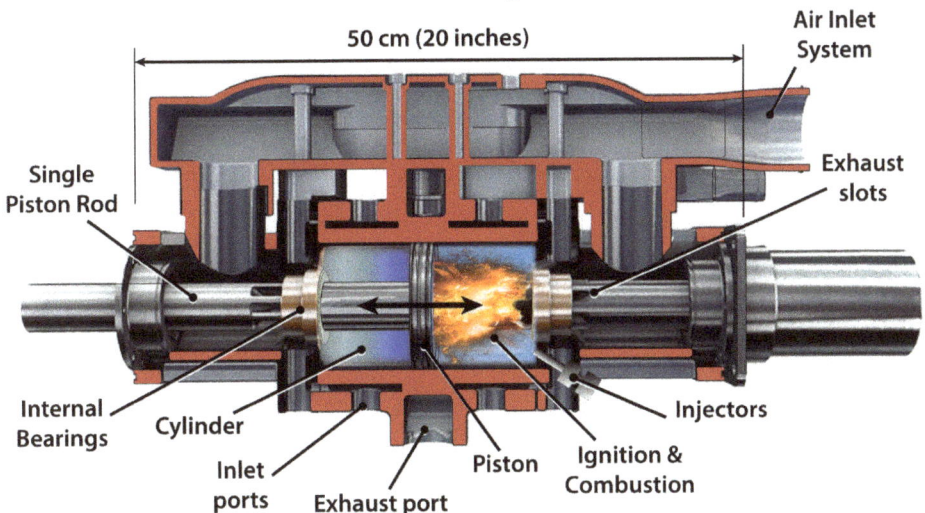

The Aquarius Two-Sided Free Piston Linear Engine (FPLE). Weighing just 10kg and with only one moving part, the piston.

LiquidPiston's hybrid-electric engine

Another exciting development is by Liquidpiston[1] who have designed a new version of the rotary engine configuration called the X-Engine that runs particularly well on hydrogen. It's 30 per cent smaller than a conventional, comparable four-stroke piston engine. This revolutionary new engine weighs just 18kg (about 40 pounds) and puts out 5kW. It has only three spark plugs, meaning that combustion occurs three times during a cycle, making it very efficient. It is also very easy to manufacture due to its incredibly low weight and lack of moving parts.

You may have heard of the Wankel rotary engine as fitted to some Japanese cars in the '70s, which was revolutionary but fraught with problems mainly due to the carbon content. This new rotary engine holds a great deal of promise.

X Mini 70cc 4-stroke Rotary SI engine

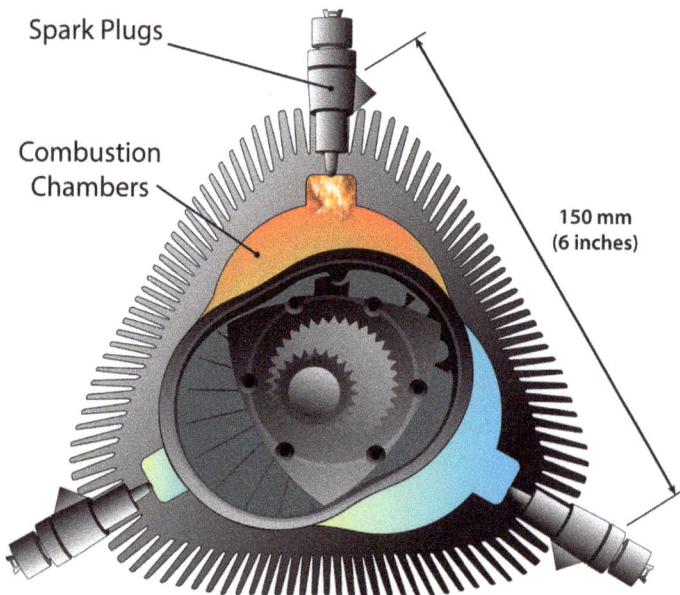

The new lightweight, rotary engine, called the X-Engine by Liquidpiston[1]

1. Liquidpiston: https://www.liquidpiston.com/

Although these two engines show a lot of promise, they are very small and unsuitable for use in large vehicles. They do however demonstrate that using hydrogen in an engine is entirely feasible, but the question still remains how best to solve the supply and storage problem. As I will explain in later chapters, a far better way is to store water and then split the water into its two gases hydrogen and oxygen only when they are needed. Literally prior to entry into the engine's cylinders. It would be very simple to convert a conventional petrol- or diesel-powered vehicle to run on hydrogen from water.

Generators

A generator is basically a static engine (although often with wheels to make it easy to move about), intended to produce 110v - 240v AC electricity, using petrol, diesel or possibly LPG gas. Generators are extremely useful and come in all shapes and sizes. From the suitcase-type or larger portable types, used by DIY'ers to run power tools, perhaps in a workshop without electricity, to farmers who need power out in a field, emergency services or road workers who need equipment and floodlights and temporary traffic lights, to campers who need home appliances and events where musicians need stage equipment, they are in use everywhere. The list goes on and on, including larger versions, perhaps still mobile but mounted on trucks or trailers for fairgrounds or emergency power for hospitals or disaster zones. Generators are also used on vehicles for supplementary power, from refrigerated trucks to yachts for navigation equipment, lighting and bilge pumps, to trains for heating and lighting.

Just imagine if all these important applications could be provided using just water. There wouldn't be fuel shortages any longer and the often 'prohibitive costs' of using petrol or diesel would be a thing of the past.

Already there are gas-powered generators running on LPG gas and even kits available to convert from petrol or diesel to LPG operation. A generator is much easier to convert to hydrogen than a vehicle engine. The main thing to remember is that when hydrogen is used, small adjustments in the timing need to be done, since the ignition of hydrogen is virtually instantaneous, unlike petrol or diesel. There is something called 'vacuum advance and retard' where the spark, intended to ignite the fuel is delayed so that the piston is in the correct position (beyond TDC 'top dead centre') for the down-stroke, otherwise the gas will explode too soon and will push the piston the wrong direction). This goes for vehicle engines too, of course.

Hydrogen refuelling stations.

Converting all our transport systems to hydrogen would be a huge issue due to the fact that currently there is no infrastructure. Hydrogen refuelling stations would need to generate and distribute hydrogen safely across the country. Sourcing the hydrogen from water is a whole different proposition. Not only that, but the process of electrolysis can actually purify or desalinate water. (See Stephen Meyer's patents for freshwater processing and hydrogen refuelling[1] in chapter 13).

Hydrogen on-demand from water could solve all the problems of supplying and storing hydrogen. Firstly, you only need to source and store plain water, creating what you need on-demand and using it immediately. So no gas is stored. Secondly, water expands an incredible 1860 times (1 litre of water becomes 1860 litres of gas), when converted to hydroxy gas and contains nearly 3 times the energy output of crude oil.

Benefits of hydrogen use in engines

- Improvement of fuel economy
- Zero emissions
- Reduced wear prolongs engine life
- Greater torque with smoother, stronger acceleration
- Increased engine power
- Zero carbon deposits resulting in cleaner lubrication
- 3 times as much energy per Kg in hydrogen than petrol or diesel

(Hydrogen has the highest energy content of any common fuel by weight being about three times more than gasoline).

---

1. Stephen Meyer, MLS-Hydroxyl Filling Station (MLS-HFS)
   https://patents.google.com/patent/US20050246059A1/en

Jet turbine car

The unlikely-sounding Jet turbine car[1] (also known as 'gas turbine'), was featured on James May's 'Cars of the people, BBC 2, episode 6', in the UK. I'm mentioning it here for a couple of reasons, one, because jet engines can be run very easily on hydrogen and a surprising range of other fuels and two, because a gas turbine is actually a very simple engine, with few moving parts. The Jet turbine car was made by Chrysler from 1963-1964 in Detroit, USA. Only about 55 were made and most were destroyed. Was it just impractical? One issue was that it required unleaded fuel as leaded fuel damaged the turbines, which wasn't widely available at the time.

So why have I decided to include a reference to a jet-powered car in this book? Well, fuel sourced from water can be used in every application that fossil fuels are currently used in. That goes for jet (gas turbine) engines too. Most people know that jet engines are used for planes but they have also been used in situations as diverse as go-karts and bicycles to auxiliary power supplies and, as here, even in cars.

1963 Chrysler Turbine in Hershey PA. Photograph by CZmarlin — Christopher Ziemnowicz, taken at the AACA show in October 1999

---

1. Chrysler's ill-fated Turbine program - Hagerty Media: https://bit.ly/3sCymxb

Aircraft

Currently we have climate protesters trying to persuade people to cut down on their flying, as it is such a massive contributor to global pollution and environmentally damaging emissions. Some protestors have even gone to the extremes of gluing themselves to the top of planes to cause maximum disruption. Having said that, the aviation industry is already looking at liquid hydrogen as an ideal fuel for aircraft and modified gas-turbine engines are already undergoing testing. (For example Airbus' ZEROe concept aircraft[1].

These smaller, lighter, simpler engines would be ideal for aircraft. In fact, the X-Engine has already been used in drones using other fuels. However, normally, the storage of hydrogen would be a huge problem, as very robust, pressurised and insulated tanks are heavy and the dangers of storing a highly flammable gas like hydrogen are also an issue, but sourcing the hydrogen from water on-demand would be an ideal solution. Stan Meyer, the designer of a water-fuelled car (as we shall see in chapter 11), has shown a design utilising water-splitting to release hydrogen for use within a jet engine. A jet engine is a relatively simple device with few moving parts although fine tolerances are required that are critical for such a high-revving engine with extreme heat and pressure.

One of the biggest advantages of using water to source hydrogen for aeroplanes, is that there is three times as much energy in water than in any fossil fuel, meaning that a plane would have a much greater range for the amount of fuel that it is carrying, not only zero-carbon, but also if the hydrogen is released on-demand, there is no fire risk either. As discussed elsewhere in this book, the US Navy has already converted seawater to jet fuel for its carrier aircraft[2, 3], which they tried to keep a secret. In fact, some online articles on the subject that I have previously bookmarked, are no longer available.

Hydrogen, along with oxygen is of course already used as rocket fuel. NASA's Space Shuttle, for example, carried tanks of Liquid Hydrogen and Liquid Oxygen as a propellant. It's a perfect combination as Oxygen acts as an oxidiser as there isn't any in space.

1. Airbus' ZEROe concept aircraft: https://bit.ly/3t8Y8cR
2. US Navy turns seawater into jet fuel: https://www.youtube.com/watch?v=Fcc-cTCVY64
3. US Navy Develops 'Game-Changer' Technology To Turn Seawater Into Fuel: https://bit.ly/3zbdzoE

Turning hydrogen into electricity

Fuel cells generate electricity by a chemical reaction (reverse electrolysis). Hydrogen (Usually from an on-board storage tank) and oxygen (from the ambient air), cause a chemical reaction to take place between the positively charged anode and negatively charged cathode. The chemical reaction strips the hydrogen atoms of electrons making them 'ionised', therefore carrying a positive charge which provides the electric current. This reaction recombines hydrogen and oxygen to form water vapour, as well as heat and an electric current.

Hybrid vehicles use these 'Fuel Cells' which turn hydrogen into electricity which then powers electric motors. These then give traction to the vehicle in much the same way as any electric vehicle, but with the added advantage of a considerably increased range. This is not by any means a new technology. It was first demonstrated by Sir William Grove in 1839, who called it a 'gas battery'. Since then, it remained just a curiosity until NASA (The National Aeronautics and Space Administration) needed a solution to provide a reliable power source capable of working for extended periods.

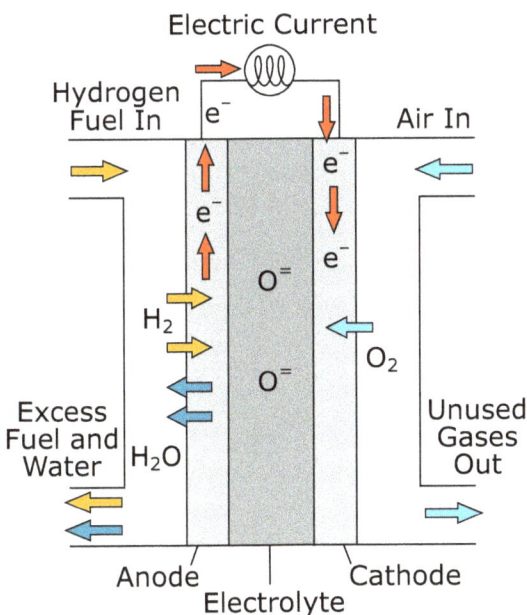

The hydrogen has to be sourced from somewhere as discussed earlier, possibly from alternative energy, but the tanks must be replenished.

With hydrogen provided from water on-demand, it would solve the problem of carrying highly volatile hydrogen in heavy tanks, with only stored water as a fuel source.

Diagram of a solid oxide fuel cell by Sakurambo
(Public domain - Wikipedia)

So hydrogen is the simplest chemical element, and the cosmos is full of it. It burns with an almost invisible flame and leaves no pollution.

We need carbon for life. But we do not need carbon to run our vehicles or power our electricity power stations. We have to stop digging up carbon-contaminating fossil fuels, and revert to using the simplest of the created chemical elements - Hydrogen.

Regardless of the cost to all the fossil fuel industries, NOW is the time to change. This struggling world can no longer tolerate more and more fossil-fuel, carbon-laden, atmosphere-polluting injections of poisoned misery.

Next we'll take a closer look at carbon. Know thy enemy! (However it does have some redeeming features, read on).

5

Carbon

World governments are now actively engaged in attempts to transform the global energy system from one where fossil fuels are dominant to one of clean technology - known as de-carbonisation - a critical step in meeting established climate goals.

The world is urgently looking for carbon-free solutions, which is ironic as we ourselves are carbon-based lifeforms. Carbon has always been part of our lives and it has some extremely useful properties, but it's really too good to burn.

Carbon is life.

Let's have a look at what carbon actually is. Carbon is life's essential component and is everywhere in our world, in fact, we ourselves are made of about 18% carbon. From an early age, we use pencils, made of carbon in the form of graphite. From the burnt wood of campfires to the charcoal of our family barbecues, we have most probably all handled carbon. It's messy!

Graphite is extremely soft and malleable, and at the other end of the scale, there's diamond, one of the hardest known substances, which is also made of pure carbon, which has been crushed underground at extreme temperatures and pressures.

The complex biological molecules in all living things are classified as organic molecules because they contain carbon. It is these that are broken down from decaying ancient forests that become what we know as fossil fuels.

Much of the fossil-fuel material comes from vast swamps and forests of about 360 to 300 million years ago, which is called the 'carboniferous' period.

Diamonds are also pure carbon that has been crushed under intense heat and pressure to become the hardest known substance. It's incredible to think that diamonds are made of the same material that we are, such is the magic of this world.

You too are a carbon life-form

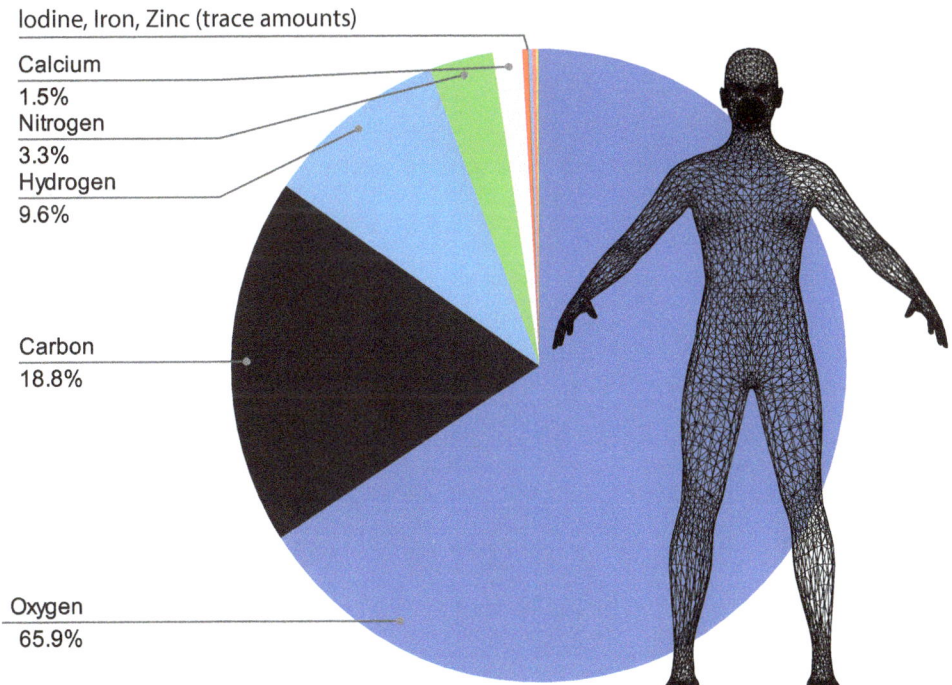

Iodine, Iron, Zinc (trace amounts)

Calcium
1.5%
Nitrogen
3.3%
Hydrogen
9.6%

Carbon
18.8%

Oxygen
65.9%

Human body ingredients by percentage.

(Pie chart made in Google sheets by the author).

Human body ingredients

1. OXYGEN 65.0% Critical to the conversion of food into energy.

2. CARBON 18.5% The so-called backbone of the building blocks of the body and a key part of other important compounds, such as testosterone and oestrogen.

3. HYDROGEN 9.5% Helps transport nutrients, remove wastes and regulates body temperature. Also plays an important role in energy production.

4. NITROGEN 3.3% Found in amino acids, the building blocks of proteins are an essential part of the nucleic acids that constitute DNA.

5. Plus Other Key Elements (Calcium 1.5% Phosphorus 1.0% Potassium 0.4% Sulphur 0.3% Chlorine 0.2% Sodium 0.2% Magnesium 0.1% Iron (trace amount) Zinc (trace amount)).

Carbon fibre

With the possibility of weaving the long molecular strings of carbon into a type of cloth called carbon fibre and sealed in a coating of epoxy resin, this advanced material can form a solid structure that possesses properties of extreme lightweight and strength. It's now often used in boats, formula one cars and aeroplanes where weight particularly matters.

You may think that carbon fibre (or 'Fiber' in the USA), is a fairly new development, but it was used in light bulbs as long ago as 1860 by Joseph Swan and Thomas Edison. The high potential strength of carbon fibre was realised in 1963 at the Royal Aircraft Establishment at Farnborough, Hampshire, UK and subsequently patented by the UK Ministry of Defence.

Carbon fibre cloth is very soft and brittle. Duboyong, CC BY-SA 4.0 https://creativecommons.org/licenses/by-sa/4.0, via Wikimedia Commons

Graphite

Graphite is quite a soft malleable material, known for its use in pencils and batteries. It is highly conductive of electricity and heat, so is very useful in electronics.

Graphene

Graphene is a relatively new super-material that is essentially an extremely thin layer of carbon, only about one molecule thick but has some extraordinary properties. Its use is being explored for a wide range of applications from anti-corrosion paint to flexible screens for wearable tech and biomedical uses.

Carbon zero?

So, we are all looking to go 'carbon zero'. That's quite an irony considering that we ourselves are essentially carbon-based life forms. When we talk about fossil fuels, we are literally talking about digging up ancient life forms, the trees of ancient forests, millions of years old and animals, both of which are also carbon-based life forms. Coal, oil and gas are all composed of the rotten remains of ancient plants and animals. Does it not strike you as a bit odd that we should do this just to recover the attached hydrogen?

Since man first discovered fire, we have been burning carbon in the form of wood and charcoal, leading on to using charcoal and coal to make metal workable and reduce limestone rock for building mortar and fertiliser. In the early industrial revolution, we found that we could make steam as a form of power to drive machinery, filthy coal being the ideal material to create enough consistent heat. And then later, we realised that thick, glutinous black crude oil could be refined down to produce a range of fuels to light lamps and power engines. All the time, what we perhaps didn't realise (and many still don't appreciate), is that through all these processes *it's the hydrogen component within fossil fuels that provides most of the energy.* (Water also contains that same energy, but of course, we can't simply burn it, as it has the unfortunate habit of evaporating before reaching its flashpoint).

When we burn fossil fuels, the hydrogen burns well but the carbon content less so, leaving unburned residue and filling the air with dangerous microscopic particulates and gases. It's sobering to think that we are still using Victorian technology, such as coal-fired steam-powered generators to provide the bulk of our electricity, even to charge up many of the new breed of electric cars that the manufacturers and governments claim to be greener than green.

So, carbon, in the form of fossil fuels has had its place, but at great cost to our health and the health of our planet.

So carbon is very useful and continues to play a part in modern material science. One way that it has served humanity is as a carrier of hydrogen. Let's look at those hydrocarbons in the next chapter.

6
Fossil fuels

Petrol | Avgas | Diesel | Benzine | Paraffin | Butane | Propane | Coal | LPG | Ethane | Octane | Hexane

The world is trying to be carbon-free. That means giving up our reliance on fossil fuels or 'hydrocarbons'. What exactly are they? How are they formed and more importantly, how are they refined into all the various fuels that we currently use?

'Hydrocarbons'

Oil is only useful because of the hydrogen content!

Hydrocarbons - fossil fuels

Fossil fuels or 'hydrocarbons' as they are known are called that because they are organic molecules mainly composed of hydrogen and carbon. We don't want or need the carbon, which is not only disastrous for our health and the environment, but it burns too slowly and clogs up all the fine machinery making it even more inefficient. Oil refining takes a vast amount of energy to extract, refine and transport and costs us dearly, but there is no need for it, as we have a massive amount of hydrogen in the form of water.

We often use hydrocarbons in our daily lives: for instance, the propane in a gas grill and the butane in a lighter are both hydrocarbons. They make good fuels because their covalent bonds store a large amount of energy, which is released when the molecules are burned (i.e. when they react with oxygen to form carbon dioxide and water).

Greenwashing

Shell, the largest company in the world in 2013, is also one of the world's biggest polluters. They are currently undergoing a process of 'greenwashing', which means they are attempting to appear environmentally friendly by advertising their adoption of green solutions. Literally, adverts have recently appeared showing wind turbines, solar energy electric vehicle top-up points, and other projects that they are funding. However, the truth is that they are spending a fraction of the billions they are putting into finding new oil fields. I think the same thing is happening with Saudi Arabia. Australia, which is committed to being carbon neutral by 2030 is happy to continue exporting coal, one of the biggest polluters.

Shell also has 1000 lawyers fighting their corner as well as unlimited resources. A real force to be reckoned with!

From seashells to fossil fuels

Have you ever wondered about the Seashell logo and name of Shell Oil Co? Well, in 1833 a London antique dealer called Marcus Samuel discovered that there was a lively market for oriental seashells, as they were popular for their use in interior design. Such was the demand that he began importing the shells from the Far East. This was the early beginnings of one of the world's leading energy companies. Marcus Samuel senior died in 1870 and passed the business on to his two sons, Marcus junior and Samuel, who began to expand it. Along with shells they also started carrying barrels of oil. When the trade for oil grew, in the 1880s, leaky barrels were not ideal, so they commissioned a fleet of steamers especially designed to carry oil and the rest is history. Today it is said that roughly half the vessels currently sailing around the world are carrying petrochemicals in one form or another.

Too good to burn!

Farm fertilisers, fabrics, almost all plastics and thousands of other vital and daily-use products including soap also come from crude oil. It has the ability to create long molecular chains that give special properties to anything we make from it, such as plastics and carbon fibre. From nylon clothing to children's plastic toys, from strong, light, plastic bags to durable parts for a million different purposes, carbon has many different forms. It's really too good to burn!

The Shell company was formed when its founder, who originally was importing sea shells saw a market for crude oil. Photograph by the author

Rock oil

The word 'petroleum' or crude oil, comes from the Latin 'petra' meaning 'rock and 'oleum' meaning 'oil'. It's a 'fossil' fuel as it's made by the decomposition of ancient organic material.

Shell Petrochemicals, the dirty truth of 'black gold' and oil billions.

I was brought up in Manchester in the UK and for a time as a teenager lived close to the huge petrochemical refinery at Partington. I could often clearly see the gas flare coming out of the top of the stack (released by pressure valves when there is too much pressure) on the horizon from my bedroom window.

What were they burning? I wondered. Why were they wasting it when heating was so expensive? It was such a futuristic sight, all illuminated 24/7, towering tanks and spaghetti of pipe work. What on earth was going on there? I got to have a much closer look when a casual work agency 'Manpower' sent me there to help unload palettes of 56 lb bags of plastic pellets from a delivery. What was the connection with petrol? I had so many questions. And why did they have a shell logo?

Of course, plastics are one of the by-products of the oil and gas industry. Another ugly environmental disaster in the making, but the very reason for their usefulness, being the reason that they don't break down naturally. For example, Nylon was a new wonder material invented around the time of my birth and promised much harder wearing clothes than those of natural fibres.

Later in my career, I was to be head hunted to help on urgent work for Cameron Iron Works, the biggest supplier of oil well-head equipment in the world. The technical publications agency had such a big contract for them that they were planning to open 24/7 and introduce shift work and maximum overtime potential to get the job done. The pay was too good to refuse too. So again I was to learn a great deal of inside information about how that aspect of the business works. The money involved was staggering! My brother-in-law also worked as a pipework designer in the oil industry. He travelled to many places in his career, from Aberdeen in Scotland to Saudi Arabia and Bahrain in the Middle East to Texas in the USA and finally Calgary in Canada. He was always paid handsomely. It seems that there's a great deal of money to be made for filthy black sludge dug up from deep underground.

When I was a child, there was a TV series I used to watch (In black and white, I might add!). It was called 'The Beverly Hillbillies' and it was a comedy about a poor family of farmers in California who discovered oil on their land and became overnight millionaires. The joke was mainly that they didn't know how to handle their newly found fortune of 'black gold'. Another thing that really brought home the power of the oil cartels was the 1970s fuel crisis. The Arab states, it seemed, had just realised the true value of their crude oil and hiked the price. The result of this was the rapid growth of cities like Dubai. Suddenly, the miles-per-gallon of our daily transport had become a key consideration. I remember clearly when I drove up to a filling station and put the first petrol into my first car. It cost me half a week's wages to fill it up! I cringed and wondered how on earth I was going to afford to run it. This led me to start investigating 'alternative technology'. It's not a new thing, wind and solar power, LPG and electric were all around in the '70s. My belief is that the government were too keen to promote their 'clean' 'cheap' nuclear power and suppress any alternatives. The dirty truth is that, in fact, electricity is only a by-product of nuclear power (from the cooling water). The main purpose of nuclear power stations is to produce plutonium for nuclear warheads, (I had been saying this for many years, but nobody believed me until Iran was accused of building nuclear power stations for that very purpose!). The requirement for plutonium diminished after the end of the cold war, when the government finally decided to consider alternative technology, 20 years too late, in my opinion.

I spent some time working at the Shell petrochemical works near my childhood home in Partington Manchester. The burning plume of excess methane gas could be seen on the horizon from my bedroom. A substantial amount of methane is released by "flaring" in this way - a major component of greenhouse gases.

The connection between seashells and oil, maybe not be at first apparent.

(It's not called a 'fossil fuel' for nothing).

Refining crude oil

Nobody seems to question how much energy they are 'putting in' to crude oil to refine it into usable fuel. In fact, it's colossal. It requires distilling at extremely high temperatures. About 400°C, depending on how refined it needs to be for different purposes and yet, somehow, everyone seems to think that the high price of oil is acceptable. The truth is that not only does it do a great deal of harm to the environment and our health, but the great majority struggle to pay for it.

Energy going into Fossil Fuels

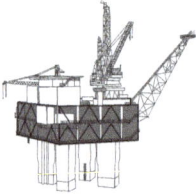

Prospecting and Drilling

Refining at up to 400°C

Transportation

Delivery

Storage and dispensing

What about the energy being put into fossil fuels? ***None of which has to go into hydrogen on-demand.*** *Plus the associated risk of environmental damage and health and safety issues of storage and delivery.*

It costs us greatly to run our cars, pay for transporting goods, heat our homes and keep our domestic electricity supply on. We don't have to drill for water, we don't have to haul it many miles in shops and trains and we don't have to buy it for rip-off prices from dodgy countries with despotic rulers. It's not dangerous to store it or use it. It doesn't damage the environment. There's too much of it in some places and where, in others, it's in short supply, we can use its power to cheaply access it from places further afield (like the Romans did!), or extract it from seawater, without the high cost of running fossil-fuel-powered pumps and desalination plants.

Refining of Fossil Fuels From Crude Oil

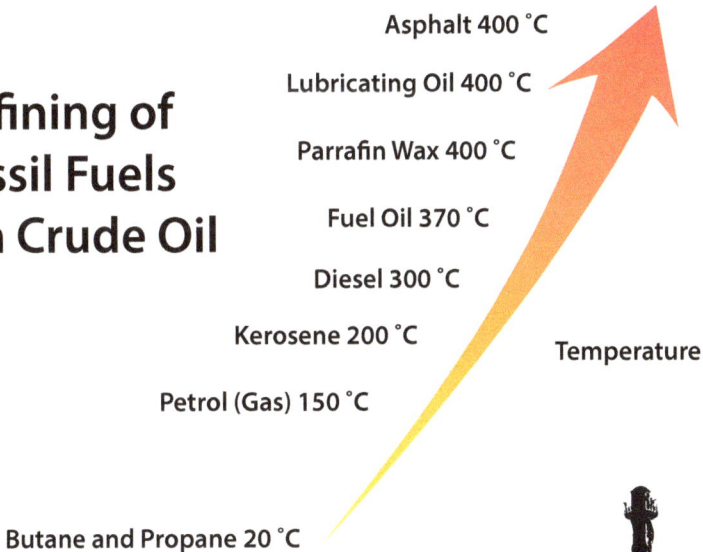

Asphalt 400 °C

Lubricating Oil 400 °C

Parrafin Wax 400 °C

Fuel Oil 370 °C

Diesel 300 °C

Kerosene 200 °C

Temperature

Petrol (Gas) 150 °C

Butane and Propane 20 °C

Many people incorrectly claim that the energy you 'put into' creating hydrogen from water simply equals the energy that you get out, meaning that it isn't worth it. However, nobody says that about refining crude oil and yet, the distillation process uses huge amounts of energy.

The carbon backbone

As you can see from the illustrations here and overleaf, all the fossil fuels contain Hydrogen and Carbon. Here are some of the most common ones that you will find burning and releasing their noxious fumes all around us in the atmosphere. They all, of course, also appear naturally. It may surprise you just how much carbon is present as the framework 'backbone' which holds the Hydrogen.

The formulae

Example
'Octane molecule'

Represents Carbon   Represents Hydrogen

$$C_8 H_{18}$$

The number of Carbon atoms   The number of Hydrogen atoms

C_8H_{18}
'Octane molecule' [1]

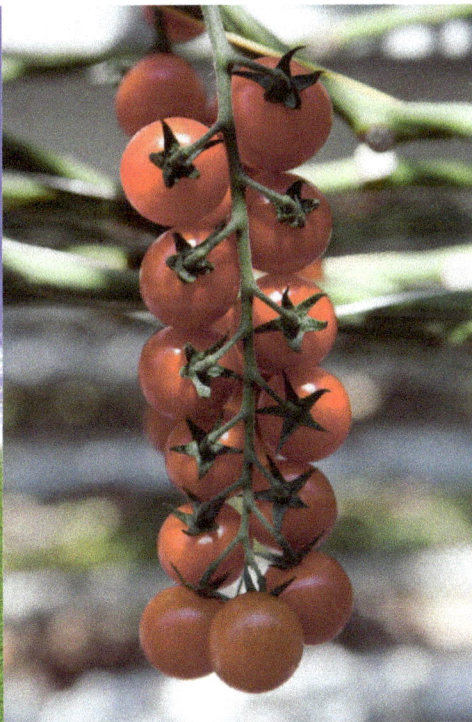

'Cherry tomatoes on a vine'
Photo by Dave Stokes [2],

Fossil fuel molecules are a bit like tomatoes growing on a vine, where the carbon strings (called the 'backbone') are like the vine and the hydrogen is like the fruit.

You wouldn't dream of eating the totally indigestible vine. In the same way, carbon is the structural component but doesn't help combustion, in fact, it impedes it by clogging everything up by coating everything with partially burned sooty deposits.

1. Rendered by the author in the open-source raytracing app 'POVRay'
2. CC BY 2.0: https://creativecommons.org/licenses/by/2.0, via Wikimedia Commons

CH$_4$

Hydrogen

Carbon

Methane

Fossil fuels are mainly hydrogen and carbon. Hydrogen has to be attached to something to be of any use, or it would simply dissipate in the air.

Carbon

Hydrogen

C$_8$ H$_{18}$

Octane

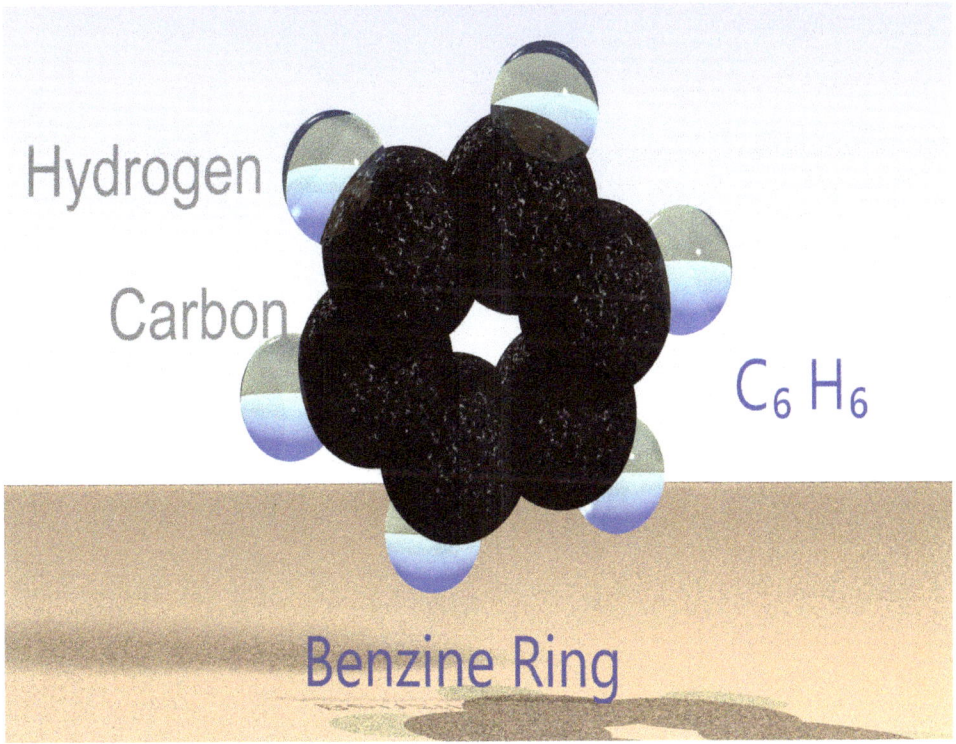

Hydrogen

Carbon

$C_6 H_6$

Benzine Ring

Illustrations Rendered by the author in the open-source raytracing app 'POVRay'

$C_6 H_{14}$

Carbon

Hydrogen

Petroleum - Hexane

Anthracite coal

Anthracite is a hard, very compressed variety of coal. With fifteen carbon atoms to every eleven of hydrogen, it has the highest carbon content, the fewest impurities, but the highest energy density of all types of coal and is considered the best quality coal.

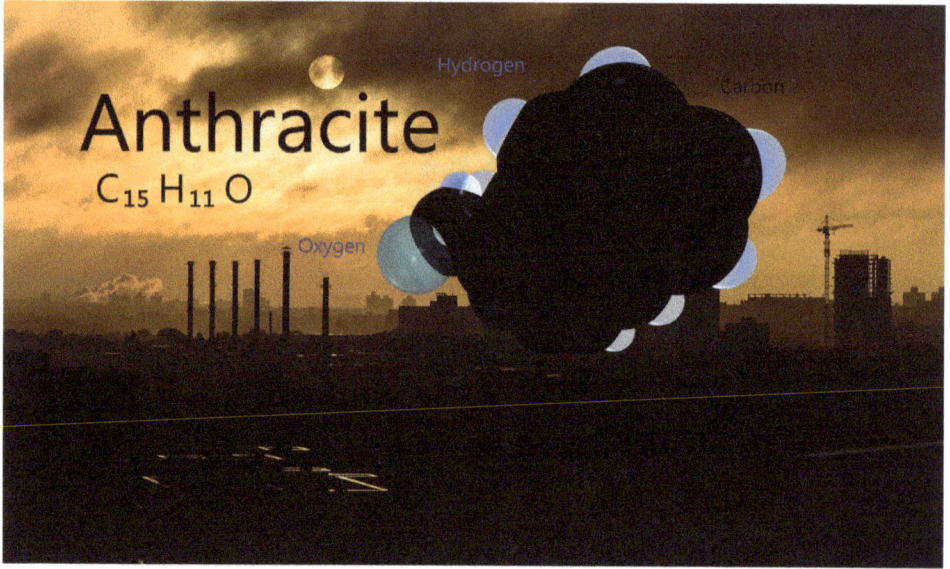

Anthracite
$C_{15} H_{11} O$

Hydrogen

Carbon

Oxygen

Rendered by the author in the open-source raytracing app 'POVRay'

Water

Hydrogen
& Oxygen

Oil

Hydrogen
& Carbon

Hydrogen is common to both water and fossil fuels.
The oxygen helps massively but the carbon is a huge problem.

Coal is the filthiest of the hydrocarbons, as it's mainly carbon. Unlike hydrogen it's quite difficult to light and burns poorly producing a lot of unburned soot and noxious gases. Burning fossil fuels such as oil, gas, and especially coal, releases carbon dioxide (CO_2) into the atmosphere, trapping heat and raising global temperatures.

Is hydrogen just an energy carrier?

I hear this being said fairly often. I think that it is a case of 'wrong thinking'. Look at the illustration on the left. If hydrogen is just an energy carrier, then the same surely applies to oil and gas too. Hydrogen is common to both in roughly the same proportions.

So hydrogen is the main combustible element of them both. If anything, water is better, as the oxygen acts as an oxidant massively aiding combustion, whereas carbon is just the opposite, a huge problem.

The reason people say that hydrogen is an 'energy carrier' is because they think it takes too much energy to obtain it, but does it? **<u>Oil certainly does!</u>**

So there is considerably more power in the form of hydrogen from water than in an equivalent quantity of any fossil fuel, but we need a novel way to release it efficiently. That novel way of efficiently releasing the combustible gas is by subjecting it to a resonant frequency. Let's look at this surprisingly familiar topic.

7

Resonance

Everything in the universe is vibration.

Albert Einstein

If you want to find the secrets of the universe,
think in terms of energy, frequency and vibration.

Nikola Tesla

We say we 'resonate' with someone when
there's a powerful connection between us.

Resonance, or vibration of particles, is a major part of how the universe works. From the sounds in your ears to ocean waves, from radio and TV waves to cooking with microwaves, it's something we experience all around us on a daily basis. I'm going to get a little bit technical to explain how it can be used to release hydrogen and oxygen from water.

Splitting water using its resonant frequency

In this section, I'm going to introduce what I believe is the main secret to the efficient use of water as a fuel, namely the idea of using resonance to release hydrogen and oxygen from water efficiently. It's well known that an opera singer can break a glass by singing at the right pitch, as long as it matches the 'resonant frequency' of the glass (yes, a musical note is a resonant frequency). The glass vibrates in unison with the musical note until its integrity is compromised and it shatters. That's what we need to do here, find the correct pitch that resonates with the water molecules and shakes them, subjecting the oxygen and hydrogen atoms to oscillating electric and magnetic fields.

So where did the idea of using a resonant frequency to split water molecules come from? Many people come across the work of the American Stanley Meyer and his twin brother Stephen who together successfully created a hydrogen-powered, water fuelled VW beach buggy as a fully-working prototype. Their car used the principle of resonance to efficiently split the water molecules, but that's not where I believe the idea came from. In my many years of research, I traced the idea back to an American-born son of a Croatian immigrant, by the name of Andrija Puharić, Doctor of Medicine (1947), who after discovering the secret in a paper by Nicola Tesla, published his patent "Method and Apparatus for Splitting Water Molecules[1]."in 1983.

I'll go into more detail about Puharić and the Meyer twins in following chapters. It's evident by looking at social media posts, that many people focus on the Meyer twin's car, but Stanley Meyer was planning to roll out a retrofit kit into mass production, which could convert **any** internal combustion engine to hydrogen on-demand using these water-splitting techniques. Also, I'll show you how after the sad demise of Stanley Meyer (during a meeting with potential investors at a restaurant, he ran outside crying "They poisoned me!", more about that later), his brother Stephen went on to improve the process by revisiting Puharić's work, culminating in a patent to produce hydrogen filling stations[2] using the very same technology but with his own ground-breaking innovations.

1. Puharić's Patent: https://patents.google.com/patent/US4394230A/en?
2. Stephen Meyer's Patent: https://patents.google.com/patent/US20050246059A1/en

Good vibrations (resonance is life).

The word resonance itself is from the Latin 'resonantia' meaning 'echo' from 'resonare', or 'resound'. It's when a sympathetic vibration occurs in one object in response to the same vibration in another. For example, a 'sound' of one object vibrating, such as a guitar string causes a vibration in your ear enabling you to hear it. (Also 'sympathetic resonance' is observed in musical instruments, for example, when one string vibrates and produces sound after a different one is struck). In an electrical circuit, it can be the play between magnetic and electric fields as one builds and then collapses in favour of the other. That's how electric currents are generated or induced and how electric motors work. *(Faraday's electrolysis as taught in schools and colleges across the world uses a current with 'zero' frequency, which will never produce resonance).*

Resonant Frequency

An opera singer can break a glass by singing a note at the exact resonant frequency of the glass. (also possible with an audio amplifier and speaker)., see the YouTube video[1]

1. Best Demonstration of Resonance - MIT professor demonstrates how glass breaks due to forced resonance https://www.youtube.com/watch?v=pyBGSMxPSG0

> ### RESONANCE:
>
> *The word itself is from the Latin 'resonantia' meaning 'echo' from 'resonare', or 'resound'.*
>
> *It's when a sympathetic vibration occurs in one object in response to the same vibration in another.*

The experiment shown below uses two adjacent identical tuning forks to demonstrate sympathetic resonance. The fork on the left is being struck by a rubberized mallet. Simultaneously, the other fork oscillates at the exact same frequency due to acoustic resonance between them. A sound wave passing through the air sets up vibrations at the corresponding resonant frequency in the adjacent one. In this case, it's slightly above an 'A' on the music scale around 460 Hz (cycles per second, a cycle being one complete wave of alternating current, voltage or sound. etc.).

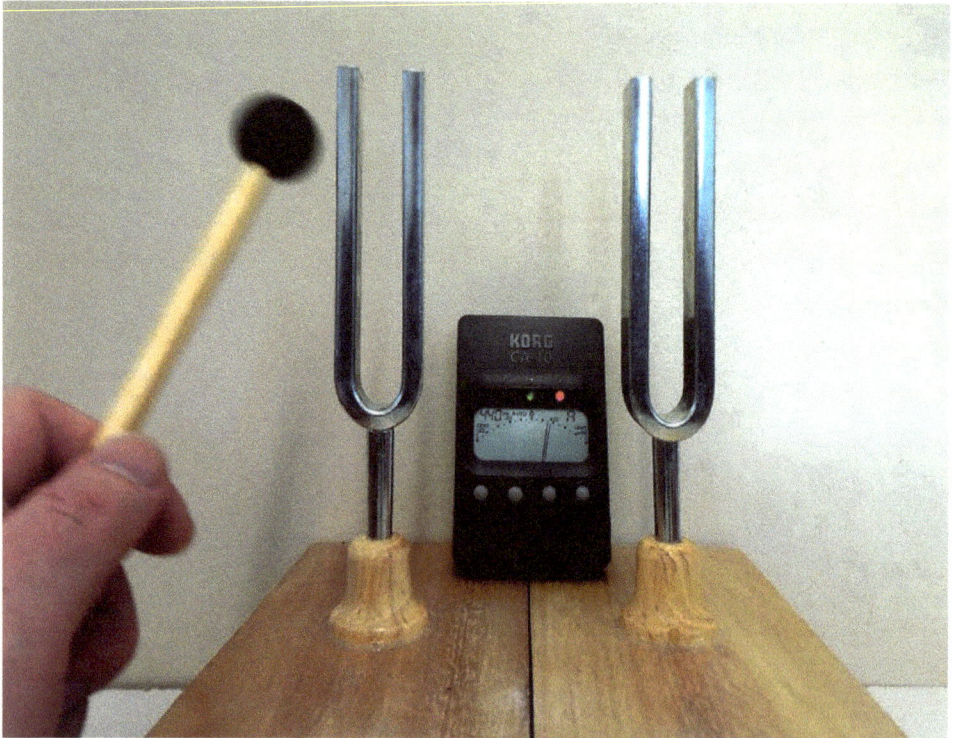

Experiment demonstrating sympathetic resonance, using two identical tuning forks. (Photograph by the Author).

Alternating Waveform

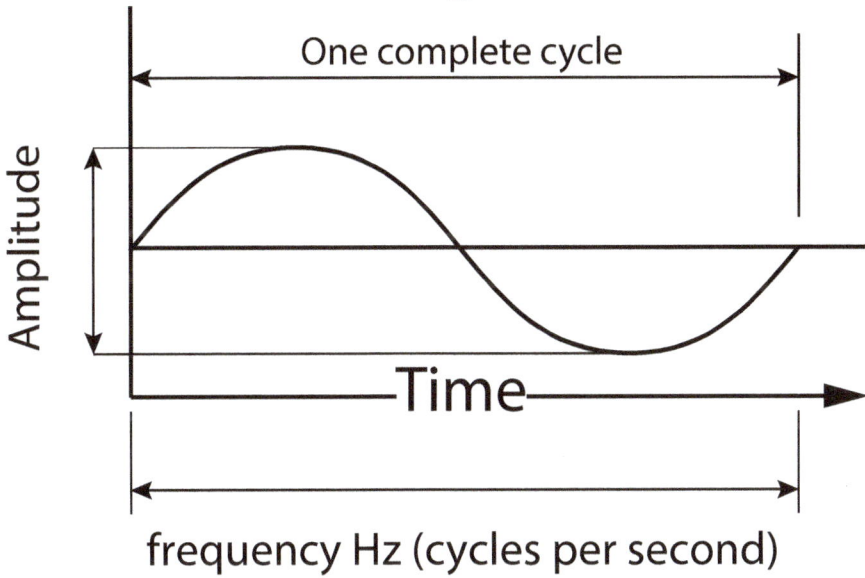

One complete cycle

Amplitude

Time

frequency Hz (cycles per second)

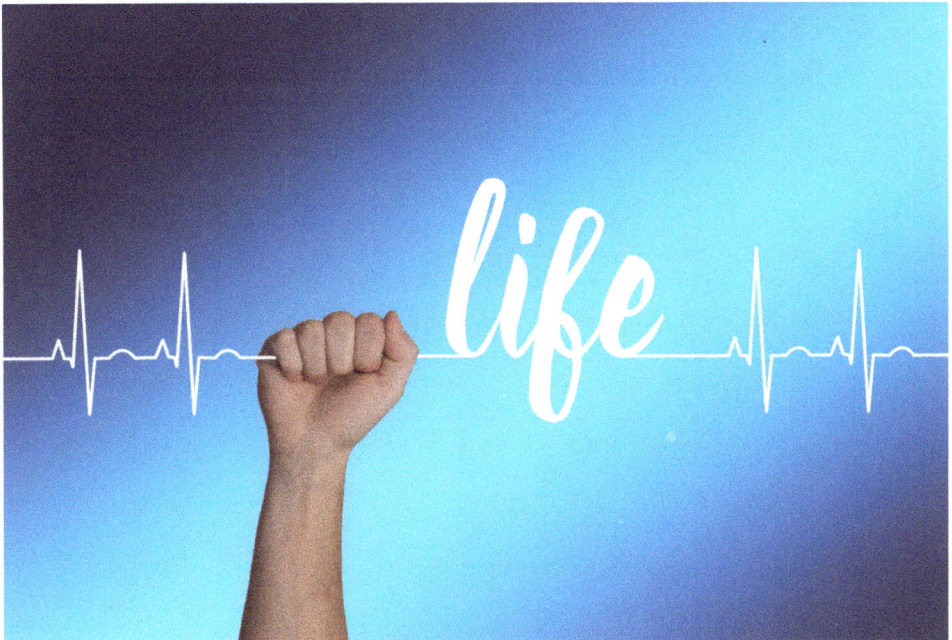

Heartbeat: Life itself is a pulsed frequency

To open a locked door or safe

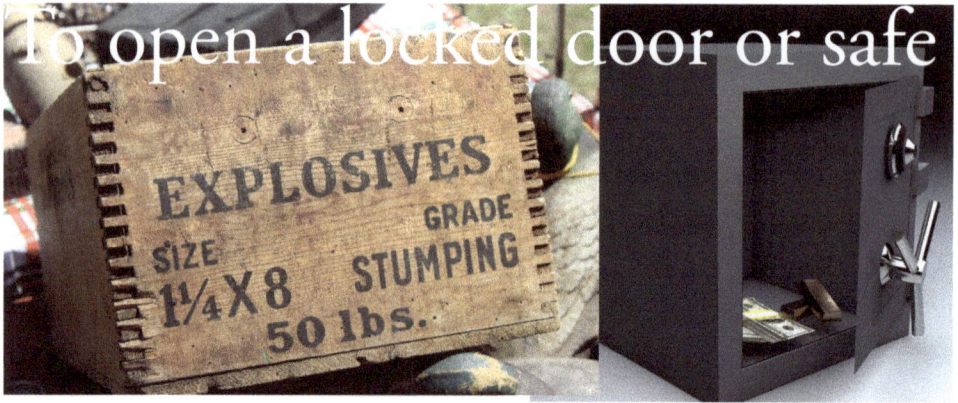

You can use **brute force**, but it uses a lot of energy, so is highly **inefficient** and risks damaging the goods...

...but if you know how, and have the key, you can unlock things with a **lot less energy**!

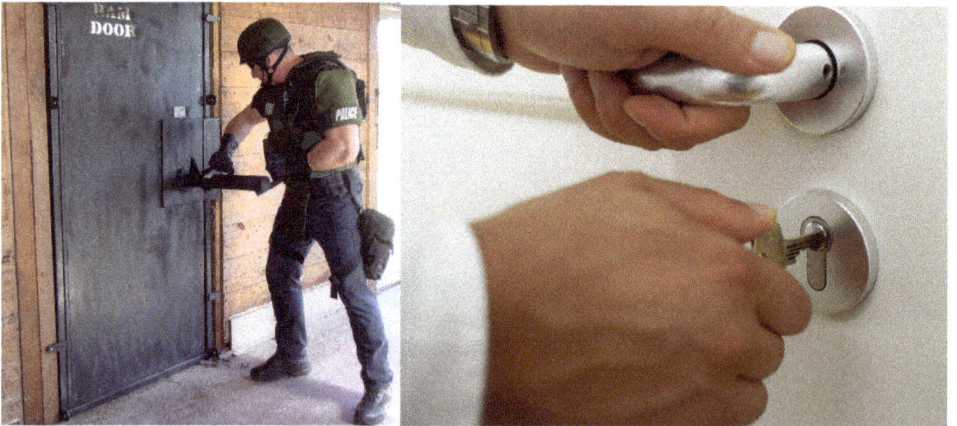

It is exactly the same with fracturing water molecules, we don't need much energy, just the right frequency!

The resonant frequency

You may think that what you learned at school, college or even University is the final word, but that can be far from the truth, especially when it comes to anything to do with magnetism, electricity and molecular physics. Discoveries that were made by great minds of the past can even be overlooked, misunderstood, or disbelieved. This seems particularly true of the subject of splitting water to produce hydrogen, in part, because much of what is happening is either microscopic in scale or even invisible, such as electricity or magnetism. We can often only study the effects of some process and then assume we know what we are looking at. For example, we have only recently been able to photograph actual atoms and atoms are not solid at all. They are more like bundles of energy and often seem to change their behaviour depending on how we are trying to observe them. If you use an electron microscope to view something on such a small scale, you are actually bombarding it with electrons in the process. In the case of atoms themselves, that's bound to change their behaviour. So, we need to keep a very open mind. Much of what we've learned is based on someone's theory or opinion and often there are arguments even between great minds until something is proven, disproven or a better theory comes along.

The public has a distorted view of science because children are taught in school that science is a collection of firmly established truths. In fact, science is not a collection of truth. It is a continuing exploration of mysteries.

Freeman Dyson

(English-American theoretical physicist and mathematician)

On the basis of all this, let no scientist tell you that using resonance is *pseudoscience* or a grey area! I'm just asking that a little of this vast amount of knowledge and expertise is directed towards the science of clean, zero-carbon energy or releasing hydrogen from water on demand, using a gated, pulsed, resonant frequency to break the molecular bonds between the hydrogen and oxygen. Too much to ask? (I'm so tempted to ask "Does that resonate with you?" How frequently do you hear that?). These are principles that we are very familiar with, so let's apply them here.

Sound is resonance in your ears

If you are a non-technical person, then you could be forgiven for thinking this idea of resonance sounds a little strange and perhaps not well known, but nothing could be further from the truth. I can assure you most people will use it in many ways, most of the time! You will find it in many diverse branches of science from power generation and battery chargers, to radio and TV to music and telephone, clocks (even mechanical clocks) and microwaves and anything with an electric motor.

Resonance, being the very reason waves are formed, occurs widely in nature. All sound and light is a form of resonance, indeed all vibrations are the foundation of matter itself. Firstly, electricity, for most of the needs of the country, is created using an alternating current at a frequency of around 50 cycles per second (in the UK), 240 volts and an amplitude determined by the current. In all musical instruments from vibrating strings or air tubes in wind instruments, to electronic oscillators, the same principle is at work. (The vibration determines the note and the amplitude determines the volume (That's where the word 'amplify' comes from). Radio waves are also oscillating vibrations tuned to a particular frequency. Many many electronic circuits use the principle of resonance too. A frequency generator, (which generates electronic signals with set properties such as amplitude, frequency, and wave shape) produces a precise tone, perhaps with harmonics, or complementary frequencies combined with a main carrier frequency. This could be part of an electronic musical instrument such as a synthesizer or a radio transmitter or used to test circuits.

Domestic Electricity supplies

Note that the 240v and 110v are 'average' voltages as they are constantly fluctuating!

The peak-to-peak voltage is more.

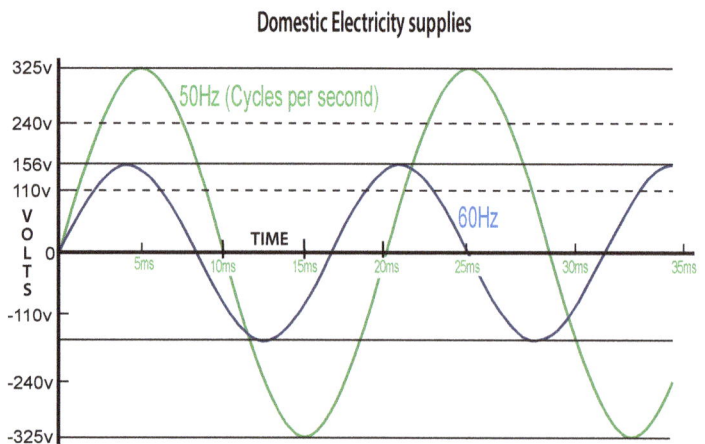

Domestic electricity supplies: UK and Europe 240v AC 50 Hz, USA 110v 60 Hz

The resonance of musical instruments

Now, I am also a musician, a passionate flute player with a collection of about twenty-five ethnic flutes, an Irish low whistle and pan pipes, as well as an electronic wind synth. I studied the 'technology of music' with the Open University, which helps my understanding of the science behind resonance. This is in fact highly relevant to the theme of this book, as I'm proposing the application of a resonant frequency which is a form of a musical note. The very same technology, *acoustic resonance,* is the study of vibrations within objects, such as the strings, and body of instruments, operating within the frequency range of human hearing. It also includes the requirement to study waveforms.

Harmonics

The main frequency is called the *'fundamental'*, or *'carrier'* frequency. *Harmonics*, created by multiples of the fundamental note, are higher frequencies which distort the waveform. They are higher vibrations, such as the same note at double the frequency, (such as 'C' and 'C' in a higher octave) understood well by most musicians. The way harmonics alter the waveform is crucial to water-splitting as we shall see later.

Most musicians are familiar with harmonics, or multiples of the fundamental frequency, achieved by touching a node at half or quarter of the string length.

The induction coil

Please allow me to get a little technical here, in order to introduce principles we need to know in order to split water into hydrogen and oxygen. You can, of course, skip it if you like, but it's worth trying to understand and I've aimed it at those who know little if anything about science. If you do happen to know a fair bit about scientific principles, please still follow, as there are uncommon principles we need to consider. The italicised terms can be Googled for more information if you are interested enough to take any principle further.

A capacitor and inductor coil on a circuit board. This combination is the heart of an oscillator circuit which features heavily in water-splitting technology. A coil holds an electric charge in its magnet field, whereas a capacitor holds it in an electric field. (When electrons move along a wire the two types of fields work at 90 degrees to each other).

Inductance

The principle of inductance is used in many applications from electricity-generating power stations to many common devices such as power transformers like the one for your phone charger or laptop. They are made by winding wire in a coil around a core, which can be a hollow, or 'air' core, to solid iron or other materials. They can be straight or in a circular donut (doughnut or donut) form. They can also be in the form of a flat coil (similar to an electric stove coil). This is also called a *pancake coil*, or *Tesla coil*.

When an electrical current is passed along a wire, it induces a magnetic field at right angles to it. Because of this, if a current is passed through a wire

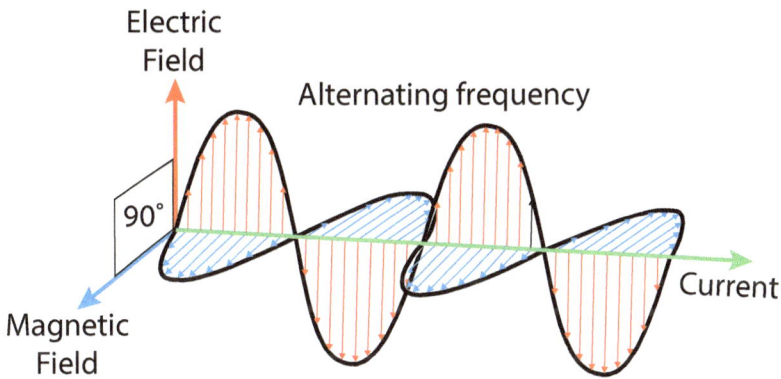

Every electric field has a magnetic field at right angles to it.

it can induce a current in another wire which is placed nearby, initiated by the magnetic field from the first wire.. Similarly, a *coil* of wire carrying a current can induce another current in a *coil* of wire surrounding it. Furthermore, if there are more turns of wire in the second coil, that induced current will be at a greater voltage. This is the principle used in those ubiquitous adaptors we plug in to charge our devices, such as mobile phones or laptops, which need a much lower voltage.

Electromagnets

If a magnet is passed across a wire, it causes a movement of electrons within the wire, the direction depending on the magnetic poles of the magnet. In other words, an electric current is induced, at right angles to the magnetic field. Also, the opposite is true, if you pass an electric current through a wire wrapped around an iron bar, it will create a temporary magnetic field in the bar, until the current is switched off again. Such a setup is called an electromagnet. If we have two wires instead of one, these effects are doubled and we can keep on winding the same wire around and around, in effect boosting the current we are creating. So that's why we have coils of wire in use in many applications. We are getting an electric current in return for the effort it takes to move the magnet, or spin the magnet within a coil. That can be done by rotating it at great speed. This is one of the main principles we need to understand here and one of the most important tools that we need to release hydrogen and oxygen from the water molecule. You can find coils using this principle absolutely everywhere in the modern world. There are coils in just about every electronic circuit, in everything from your mobile phone to your TV, from your hair dryer to your hi-fi. They are the basis of microphones and speakers and headphones and also devices such as the microwave. Coils allow voltages to be "stepped up" or "stepped down" in mains power adaptors, (also called 'transformers'), that we use to plug lower-powered accessories into the wall sockets (say for example to reduce the current from 240v AC to just 5v DC to charge a mobile phone). Indeed coils allow electricity to be produced in the first place, in the giant coils of a power station's generator. Electric motors use the opposite principle, using electricity to move magnets arranged around a spinning hub. These range from tiny ones, for example, those in a hand-held fan to the large ones powering electric vehicles.

A large, modern, electric generator.

Oscillators

An oscillator is a mechanical or electronic device that works on the principles of oscillation: a periodic fluctuation between two things based on changes in energy. Computers, clocks, watches, radios, and metal detectors are among the many devices that use oscillators. Energy charges up a magnetic field and when that collapses the energy is transferred to an electric field and vice versa. This is the principle of the oscillator. This alternate collapsing and building of the fields can be caused by an *alternating current* constantly switching direction, many times a second. Significant here, because it's an electronic oscillator that is used to create the resonant frequency which is required to split the water molecule. It works by constantly subjecting the molecular bonds to alternating fields causing constant flexing, which will ultimately break them.

The 'seesaw' tank circuit.

I want to now introduce the oscillating 'LC tank circuit' or 'tuned circuit' which oscillates at its natural *resonant frequency* and is one of the most useful circuit configurations in electronics. It is a very common circuit used in many applications particularly radio and audio systems. It's called a 'tuned circuit' a 'tank circuit' or an 'LC circuit' ('L' representing the inductance and 'C' the capacitance). (There is also the 'RLC circuit' including a resistor, which dampens the oscillations. This is known as a 'harmonic oscillator'). The resonance is provided by a coil or 'inductor' (literally a double-wound coil of wire in which one wire induces a current in the other) and a capacitor (as the name suggests, holds a quantity of electric charge). Like a child's seesaw, energy oscillates between the inductor on one side and the capacitor on the other. The inductor charges up the capacitor and when full, the capacitor's field collapses, charging the inductor once again, this cycle repeating over and over as long as power is applied.

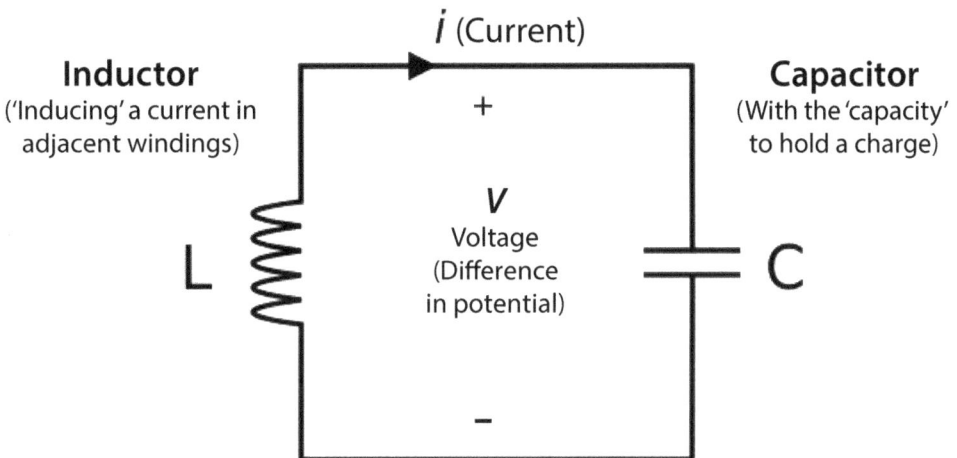

i (Current)

Inductor
('Inducing' a current in adjacent windings)

+

Capacitor
(With the 'capacity' to hold a charge)

L

V
Voltage
(Difference
in potential)

C

–

The basic LC 'Tank circuit' consisting of an inductor/capacitor combination, which oscillates at its natural resonant frequency.

A condition of resonance will be achieved when the reactance of the inductor and capacitor are equal to each other, like a seesaw when evenly balanced. Because the inductor's reactance increases with increasing frequency and the capacitor's reactance decreases with increasing frequency, there will only be one frequency where the two reactances are equal. That is the 'resonant frequency' that we are hoping to achieve. What happens when this state of resonance is achieved is astonishing.

When capacitive and inductive reactances equal each other, the total impedance **increases to infinity**, meaning that **zero current** *is drawn from the AC source!* Using a formula we can work out the exact frequency achieved with a particular inductor and capacitor pair. (You may skip this section if you don't have a head for figures. It's only included to add weight to my argument and acceptance within the scientific community, who currently are fond of claiming that all this is 'pseudoscience').

The Resonant Frequency is
$$f = 1 / (2 * \pi * \sqrt{(L * C)})$$

(Where: f is resonant frequency, L is circuit inductance & C is circuit capacitance)

The charge coil and electrolyser setup that is used in resonant frequency electrolysis, is an LC Tank system, because the electrolyser with its positive 'Anode' and opposing negative 'Cathode' acts as a capacitor. The whole thing oscillates at a certain frequency (so many cycles per second) and a certain voltage perhaps with a period of activity and a period of rest, called a *gated frequency*. The amplitude governs the amount of current running through, which in its turn determines the amount of energy being used.

*When capacitive and inductive reactances equal each other, the total impedance **increases to infinity**, meaning that **zero current** is drawn from the AC source!*

Resonant transformer

When a capacitor is connected across *one* winding of a transformer, making the winding a tuned circuit (resonant circuit) it is called a single-tuned transformer. When a capacitor is connected across *each* winding, it is called a double-tuned transformer. These resonant transformers can store oscillating electrical energy.

Isolation transformer

There's also something called an 'isolation' transformer, which uncouples the direct connection between parts of a circuit whilst still doing its work. This is sometimes necessary to safely separate, say a very low from a very high voltage. An 'optocoupler' is one way of doing this, as it uses light (as in 'optical'), so there is no actual electrical connection between them. (It's usually done using a light emitting diode and photoresistor combination).

Polyphase electricity

One thing I want to introduce here is the principle of 'polyphase' electricity. Why we are interested in polyphase systems here is that three phases or more are much more efficient. Electricity is provided to homes, usually, as single-phase, or dual-phase, but for industrial applications, where often more power is required, three phases are common. Some late model cars already have six-phase alternators to provide power to the ever-increasing electrical requirements of the modern vehicle. Stephen Meyer (see chapter 13), realised that using a double three-phase electrical generator could provide a highly efficient output of hydrogen when applied to water fuel cells.

When you use a spinning coil to produce electricity, any point on the rotor will produce an electric current moving from minimum to maximum voltage and back again as time progresses, producing the familiar sine-wave. (Possibly 50 cycles per second or 50-60 Hz, as in the UK), as discussed previously. However, you can tap a coil in many places at the same time and get more power from it. For example, if you tap the coil every 120⁰, you split the power output into three separate phases as three different sine waves. In other words, one will always be at or close to maximum voltage output at all times.

I think that it's worth mentioning here that the terms 'positive' and 'negative' when related to electricity are VERY misleading! 'Negative' does not mean an absence of electricity, it simply means that it is moving in the opposite direction. Also, it depends on the two points that you are measuring from. Voltage measurements, for example, measure the difference between the two points (called the electrical *potential*). (Think of it like a pendulum swinging 'left and right', or 'backwards and forwards').

The Tesla 'bifilar' coil

Nikola Tesla invented a special type of coil, called a *'bifilar coil'*, which he considered so important that he patented it[1], and yet has largely been overlooked. The "COIL FOR ELECTRO MAGNETS,[1]" is a flat 'pancake' parallel-wound, series-connected bifilar coil, and has some rather special properties that will be very useful to us in our quest for efficiency. Due to the unique way that it is wound, the coil can hold a greatly increased amount of energy in its electric field, and lower resonant frequency than coils wound conventionally. A bifilar wound coil maximises the voltage difference between its turns and minimises the current. Some experimenters have also found that the coil can act as an inductor AND one plate of a capacitor at the same time. *It's possible to set the bifilar coil into oscillation at its resonant frequency* using an external power source.

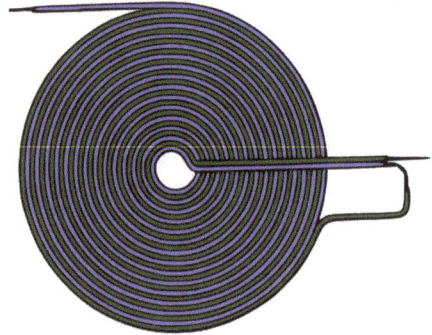

Nikola Tesla's bifilar coil (United States patent 512,340 of 1894 [1]). It has special properties due to the unusual configuration of its winding, such as increased self-capacitance. The voltage potential between any two adjacent pairs of wires is constant.

Coils are usually just one wire wrapped around a core. With this design, two wires are wound side-by-side and then the end of one wire is connected to the other. This gives an even potential across the whole coil, between any adjacent point along the wires. At resonance, the coil will naturally counter the resistive 'counter e.m.f' or 'electromotive force', which usually resists the inductance. Because of this, the coil can rapidly build its charge to a very high potential.

There are other forms of bifilar coils which are wrapped around a core in the usual way, but connected in a similar way to the pancake coil.

---

1. Coil for electro-magnets Patent US512340A:
   https://patents.google.com/patent/US512340A/en

Resonant frequency electrolysis

Obtaining hydrogen from water isn't difficult. In fact, it is so easy it can be, and is, done by schoolchildren (You may even have done it yourself). All you need to do is pass a 9v - 12v dc electric current ('direct current') through a positive and negative rod placed in the water and the hydrogen gas will form small bubbles around the negative rod (called the 'cathode', and oxygen bubbles will form around the positive rod (called the 'anode'). This can be collected by placing a suitable vessel above it. If it is so easy, you may ask why it seems to be such a problem. The answer is that the energy that you get from a few small bubbles of gas just about balances the energy that you are using in the form of an electric current to extract it. That is, however, only true if you use Faraday's non-resonant form of the experiment.

Super-efficient electrolysis

So how is this *new* 'super-efficient' electrolysis achieved? We are still applying electricity to the water, but instead of a continuous 'DC' current ('direct current'), we are going to use an 'AC' current, or 'alternating current' (just as is used in a domestic property), but it rapidly fluctuates from minimum to maximum voltage at an extremely rapid rate, maybe up to 500 times per second (50Hz). Each complete period is called a 'cycle'. (Domestic electricity in the UK is 50 cycles per second at 240v). We may also need a 'gated pulse', meaning that the voltage is 'on' and then 'off' for a period of time. (I'm deliberately keeping this in as simple terms as possible, but there are links to more in-depth technical information in the Appendices). Now, what we are aiming for is to use as little current as possible (think of a river current, the 'force' of the water, the 'frequency' as the number of waves and the 'voltage' the height of the waves). We need a high 'voltage', an intense 'spark' in fact. If you think about it, most vehicles already meet these criteria. A 'High tension' circuit of between 5,000 volts to 15,000 volts is provided by the coil and produces the ignition sparks to the fuel mix and an 'AC' circuit is provided by the spinning alternator! So, can we just use this setup to release hydrogen from water? Well no, not quite. To succeed, *we need a very precise voltage and frequency to match the resonance of the water*, a '*resonant frequency*'. Also, constant adjustments need to be made to account for the temperature, volume, water density, etc. Not at all difficult in our age of electronics.

Fracturing the water molecule

Along with using a resonant frequency, which causes the building and collapsing magnetic field within the capacitor, a water fuel cell also acts as a capacitor as it has positive and negative electrodes. This is the secret to fracturing the water molecule thereby releasing hydrogen and oxygen. **A building and collapsing magnetic field can increase the current by as much as 1000 times with no extra draw on the source**, as long as the coil is tuned to the resonant frequency. The energy stored in a magnetic field is often disregarded as it is usually wasted energy which becomes 'back emf' (electromotive force), or 'static', but in the case of a water fuel cell, this 'spark' energy is what focuses all the energy where we need it, fracturing the water molecule.

"In its simplest form, resonance occurs when an object experiences an oscillating force that's close to one of its "natural" frequencies, at which it easily oscillates. That objects have natural frequencies is one of the bedrock properties of both math and the universe. A playground swing is one familiar example: "Knock something like that around, and it will always pick out its resonant frequency automatically, Or flick a wineglass and the rim will vibrate a few hundred times per second, producing a characteristic tone as the vibrations transfer to the surrounding air."

Matt Strassler,

(particle physicist affiliated with Harvard University)

Sound vibrations

Sound is really just vibrations in the air, with which you can hear music, trains, planes, and cars, people's actions and speech. It is often measured by its frequency, which is measured in 'cycles per second' with a unit called a hertz. One Hertz (Hz) equals one vibration per second, named after Heinrich Rudolf Hertz (1857–1894), who was the first person to provide conclusive proof of the existence of electromagnetic waves.

Sound is described as a compression wave. Meaning that it travels as a force that gets transmitted through the air molecules with an initial push. These pushed molecules bump into each other.

"Quantum theory revealed that the structure of atoms, no less than the structure of symphonies, is intimately tied to resonance. Electrons bound to atoms are a little like sound waves trapped inside flutes."

"How the Physics of Resonance Shapes Reality"

"The same phenomenon by which an opera singer can shatter a wineglass also underlies the very existence of subatomic particles".

Quotations from 'Nautilus' by Ben Brubaker[1] February 9, 2022.

1. Freelance science journalist whose writing has appeared in Quanta Magazine, Scientific American and The Conversation, as well as on his blog, Measuring in Reflection. He has a PhD in physics from Yale University and conducted postdoctoral research at the University of Colorado, Boulder.
https://nautil.us/how-the-physics-of-resonance-shapes-reality-13915/

For an excellent introduction to resonance, you could do no better than this YouTube video.: Resonance and sounds of music lecture' by Walter Lewin.
https://youtu.be/f4M-6tWtkoA

Boosting the voltage

In order to split water molecules efficiently, we need to boost the voltage and reduce the current, similar to the way in which a car's electrical system has an HT (High Tension) circuit producing large voltages from its coil to create a spark that ignites the fuel/air mix. What we are most interested in here though, is the way we can massively boost a voltage using a coil with maybe a thousand turns.

One coil type of particular note is that in a vehicle's engine. The 'ignition' coil usually boosts a car's battery voltage from around 12v-14.4v to between 15,000 and 40,000 volts, required to create the spark that the fuel needs to ignite in each piston. Sometimes power transistors are used instead of coils, and there are also systems, particularly on motorcycles with two-stroke engines, that use a capacitive discharge (charge dumped by a capacitor), instead of a coil.

Diesel engines have a slightly different system, where the compression of the fuel within the cylinder ignites it, as diesel is much harder to ignite than other, more refined fuels.

A much more efficient way of heating water

(Ohmic Array Technology)

Whether it's just to boil a kettle for a hot drink or hot water for other domestic tasks like washing dishes or general cleaning, perhaps for hot baths and even Jacuzzi's, we somehow need energy to heat water. It's just another way that we use massive amounts of fossil fuel. Due to so-called fuel poverty, many people in the UK find themselves having to choose between food or heating. Due to the high cost of heating water during the current fuel crisis, a local swimming pool is faced with closure as it has sadly become uneconomical. I expect there will be many more to follow. Hot water is used to heat homes and other public buildings through the use of boilers and heat radiators. Then there's the use of steam, which we covered a little in chapter 1, (remember that most of our electricity is, in fact, produced from steam provided by coal-fired, gas-powered, or even nuclear-powered electricity generation power stations!). But guess what? There's a far more efficient way of doing it and that is, you guessed it, resonance. In chapter 11, we'll look in detail at the water-powered car built by Stan Meyer and one of the inventions that he utilised in that, to stop the water tank from freezing, was a 'steam resonator'. This used resonance to excite the water molecules heating them up. Another inventor that uses a form of resonance to heat water is Jean Christopher Dumas[1]. He uses a principle called 'Ohmic Array Technology'[2].

His device uses two bell-like metal semi-spheres, set one within the other with a very small gap. It is put into a container of water and mains electricity is applied, one pole to each of the two semi-spheres. The AC power alternates at 50Hz (50 cycles per second). This causes friction in the water molecules which rapidly heat up to boiling point. This technology has been proven to work, but somehow hardly adopted. The plates do not have to be spherical however and can, for example, be two flat plates of carbon. The correct gap is important as is the correct frequency. One company that has Patented[3] their version is Heatworks[4] and is even offering a coffee maker and a dishwasher that use the technology.

I wonder when we'll start to look seriously at these alternatives?

1. http://www.rexresearch.com/dumas/dumas.htm
2. 'The Dumas Effect - Free Energy Water Heater':
    https://www.youtube.com/watch?v=gmgDIvPWRtQ&t=23s
3. 'Liquid heating apparatus and method' Patent:
    https://patents.google.com/patent/US7742689B2/en
4. https://myheatworks.com/pages/about-the-technology

The vehicle alternator

In order to produce a resonant frequency, you will of course need something to supply the electronic signal. Andrija Puharić (as we'll see in chapter 10) used an audio amplifier and signal generator, but Stanley Meyer (chapter 11), used a modified alternator to produce the resonant waveform for his water-fueled car. This was possible because a vehicle alternator produces a sine-wave output. It does this by turning rotary mechanical motion, usually by the means of a rubber belt driven by the engine, into an alternating electrical current. This is achieved by the means of electric induction between an armature consisting of many coils of wire spinning within a stator possessing a magnetic field.

The alternator's job is to provide electrical power, not only to the engine's ignition system, fuel pump and electronic management system but also for all the electrical items from lights, cabin heater, windscreen wipers, electric windows, sound system and heater, etc.

This alternating current can also be used to provide a multi-phase, pulsed, resonant frequency to a water-fuel electrolysis system.

So, using the very familiar and abundant principle of resonance we have discovered a way to fracture water molecules efficiently.

If you managed to follow that, well done. It's a bit technical but not exactly rocket science. In the following chapters, I'm going to look at the work of several genius people who have succeeded in getting it to work. Having discovered the secret, Puharić ran a seven-litre RV (Motorhome) for thousands of miles across the USA only using hydrogen from water as a fuel. I'll Look at his work in detail, including important parts of his patents. Stanley Meyer and his twin brother were awarded patents and got a VW beach buggy to run on water alone and I'll look at several other people in subsequent chapters who have successfully released hydrogen and oxygen from water to use as a fuel using resonance.

8

Electrolysis

electro signifies the use of electricity and *lysis* is from the Greek 'to split, loosen, set free'

How to make fuel by splitting water

Electrolysis is the technique of applying an electric current to water (or other catalyst) in order to disassociate the constituent oxygen and hydrogen gases. Traditionally done with a DC (direct current), but herein proposed as an 'enhanced electrolysis', using AC (Alternating Current), with a 'resonant frequency' of a specific waveform to efficiently release the oxygen and hydrogen to use as a fuel source.

Michael Faraday's laws of electrolysis are based on his research published in 1833. Trying to use a theory from 1833 to disprove 21st-century technology is simply insane. We're talking about events at the atomic level and the electron wouldn't be discovered for another 64 years, 30 years after his death. (Discovered by English physicist J.J. Thomson in 1897).

Split, loosen and set free.

The term Electrolysis ('electro' signifies the use of electricity and 'lysis' from the Greek 'to split, loosen, set free'), was coined by Michael Faraday in the 19th Century. In other words, it is the art of adding just enough energy in order to split the two gases apart, so as to release the energy that they already possess.

We surely all know that hydrogen possesses explosive power and oxygen is usually required (as an 'oxidant') to help the burning process, producing heat, gases, and light (flames).

Making hydrogen is easy

It's not difficult to make hydrogen from water. Anyone can do it. You may even have done it yourself at school or college. The big question is whether anyone can do it efficiently enough for it to be practical. And the answer is, most definitely.

> If you want to make a little hydrogen, just pop a fully charged battery (e.g. an AA or PP9) into a glass of water that has had a little Epsom salt added to make the water conductive. The small stream of bubbles forming at the positive terminal will be oxygen and hydrogen bubbles will form at the negative terminal. Congratulations, you've just separated H_2O into H_2 gas and O_2 gas. There will be twice as much hydrogen as oxygen.

Electrolysers

Water electrolysis is performed by a device called an 'electrolyser'. (Although what most people think of as an electrolyser, Faraday called it a 'Voltameter'. He used it mainly to measure the voltage required to produce a certain quantity of gas, not the other way around).

There are various types depending on their usage and output required. The usage might be to produce hydrogen for improving an engine's fuel burn (see chapter 9 'HHO') or to produce a usable flame to use as a cutting torch.

School electrolysis

The way that electrolysis is taught in schools using only a DC current, is simply inadequate and misleading. The standard school electrolysis demonstration, based on Faraday's, uses a DC current rather than a far more efficient AC gated resonant frequency.

Due to limited school budgets, the equipment is often less than perfect too. For example, they usually have a tiny strip of platinum wire for the electrode as platinum is very expensive, and surely does more to convince people of the lack of efficiency of the process.

If the experiment is only going to demonstrate an inefficient process, why bother teaching it at all? It doesn't have to be inefficient if you do it correctly.

No wonder it's difficult to convince people that efficient splitting of water molecules into usable hydrogen and oxygen can be done. It's a bit like teaching how to build pedal cars and then claiming as a result, that cars could never be practical as a human being couldn't possibly pedal fast enough.

What we are talking about in this book is splitting the water molecule efficiently. In the name of efficiency, every part of the process counts and that means that we may need to re-address our assumptions about the science that we are using. It seems that the science we are taught at school is not a set of hard and fast rules, but often theories as they apply to phenomena that we have observed over time. When we talk about what is happening at a molecular level, it seems that we still have more questions than answers, particularly regarding electricity and magnetism, specifically fields. For example, the work of the amazing Tesla, who gave us alternating current and many other aspects of our technological world, still holds a few mysteries. What we need to succeed in efficiently splitting the water molecule is to use very little current but fluctuating resonant voltages (and to that end, we find the use of Tesla's 'bifilar coil' particularly useful).

If you are interested in electricity and magnetism, the work of people like Charles Proteus Steinmetz, (German-born American mathematician, electrical engineer and also a professor at Union College) throws some interesting light on the subject, overturning many generally accepted theories.

Wet cells and dry cells

There are two basic types of electrolysers, wet cells and dry cells. The basic electrolysis experiment uses a wet cell, in the form of a flask of water (or catalyst), with the anode and cathode electrodes sitting in the fluid. Other types of wet cells use nested pairs of tubes made of stainless steel or other metal, with the positive tube within the negative (earthed) tube (as in Stanley Meyer's patent for his 'water fuel cell'). Often they use flat plates of stainless steel. The misleadingly named 'dry' cell, on the other hand, has fluid within it, instead of the anode/cathode sitting in static fluid.

The double-tube array water-fuel cell. From the top and front views

The process of water electrolysis using the Hoffman apparatus, shown on the right, Invented by Wilhelm von Hofmann (1818–1892) in 1866, is typically used in educational institutions. An electric current (usually DC Direct Current) supplied to the two terminals, (Red for positive and black for negative) is passed through the electrodes into the water to complete the circuit, a tiny stream of gas bubbles is formed, oxygen on the anode (left-hand side) and hydrogen on the cathode (right-hand side). Water being H_2O means that twice as much hydrogen is produced than oxygen and the gas collected in the upturned test tubes. The procedure is followed by the student testing the gas by introducing a burning splint into the test tubes in turn. Hydrogen will ignite with a loud 'pop' and the oxygen will re-ignite a glowing splint.

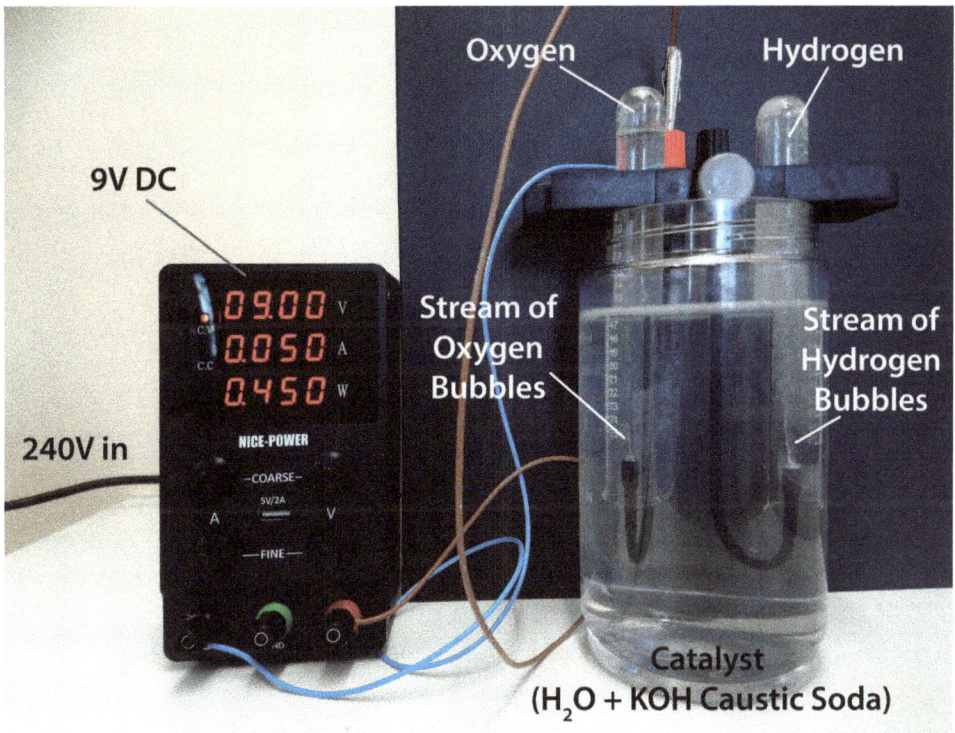

Water Electrolysis Demonstration, typical of that performed in schools and colleges the world over. With a DC voltage and tiny strips of platinum foil for the anode and cathode, it only serves to prove that the process of water-splitting is inefficient. It's misleading. It's so inefficient that it took 1 hour to collect just 1/4 litre of hydrogen. (Photograph by the Author).

Catalysts and electrolytes

Water in an electrolysis cell is often called an 'electrolyte' and may need chemicals added to make it more conductive. Which could be sodium hydroxide (NaOH) or potassium hydroxide ('KOH, also known as 'lye' or 'caustic soda'), both used in the manufacturing of soap and sold in pellets, flakes, or powders). Be aware that they are dangerously corrosive, but very little needs to be added. Dissolving salt in water also creates an electrolyte, which is why seawater is more conductive, but the salt being sodium chloride will produce toxic chlorine gas in the process. However, since a high voltage is used with 'enhanced electrolysis', it is not necessary to add anything.

The 'Dry' Cell

A modern 'Dry cell' electrolyser has such a misleading name, as it doesn't sit in a water bath like a wet cell. Instead, the water or electrolyte passes through. They are extremely easy to construct as they are made of simple flat plates, maybe square, hexagonal or octagonal. They can be made of acrylic plastic and bolted together with stainless steel bolts, which also serve as electrical connectors passing through all the plates. The plates are stacked with rubber seals in between to electrically isolate them. Plain or distilled water is used or sometimes a little catalyst is added (e.g. KOH, Potassium Hydroxide, an inorganic compound used to make soap and cleaning solutions) to make the water more conductive. Importantly, a maximum of 1.5v - 2v per cell is required. Any more than that will start to heat up the water, but not make any more gas.

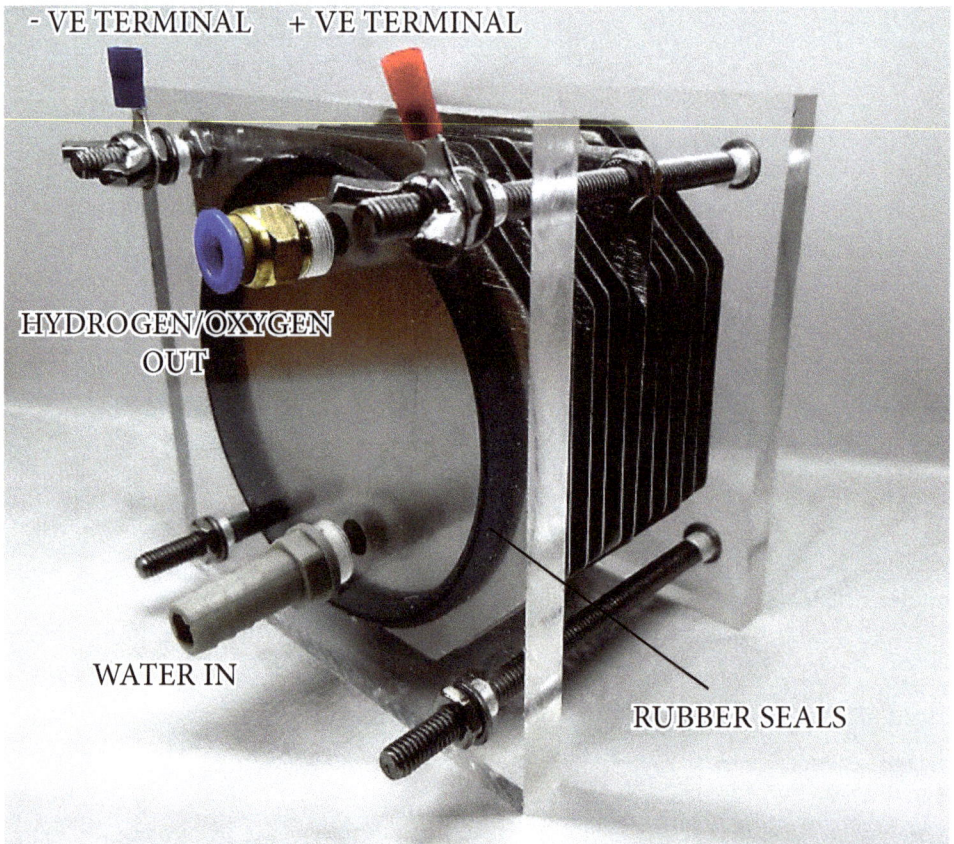

A so-called modern 'drycell' can have any number of plates producing hydroxy gas. Some have over 100. (Photograph by the Author).

A larger surface area = more efficiency!

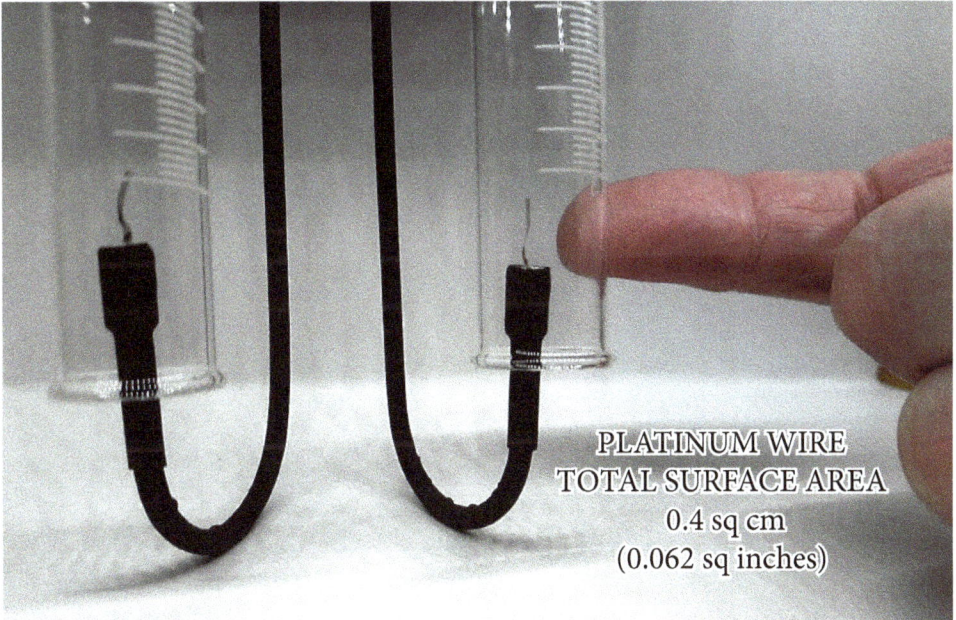

PLATINUM WIRE
TOTAL SURFACE AREA
0.4 sq cm
(0.062 sq inches)

The tiny piece of platinum wire in a typical school electrolysis setup compared to a modern HHO 'Drycell' with 11 stainless steel plates.

That's 5,500 times more area!

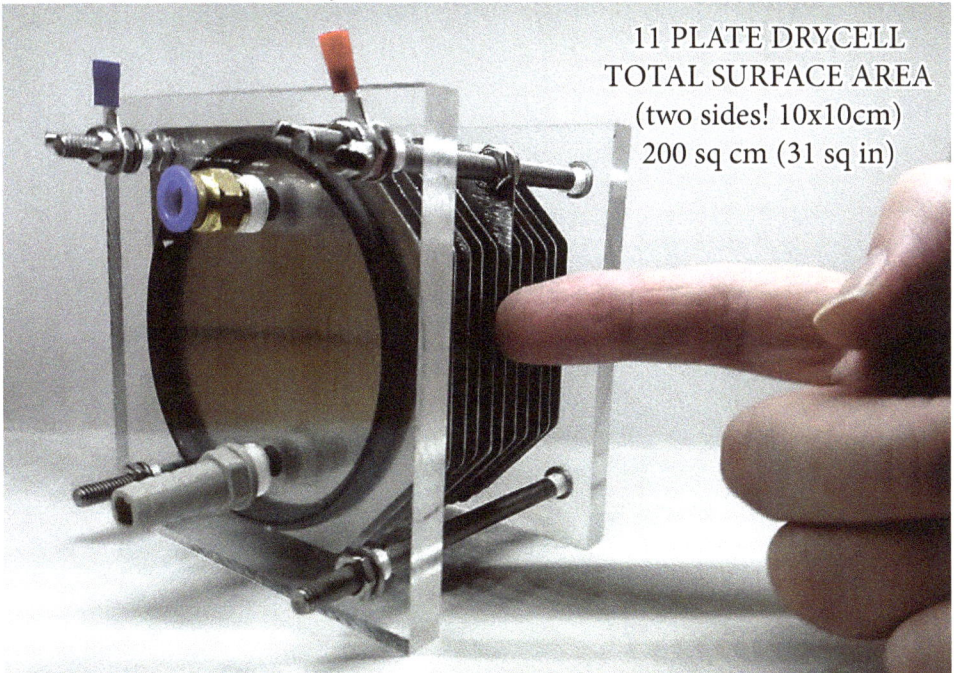

11 PLATE DRYCELL
TOTAL SURFACE AREA
(two sides! 10x10cm)
200 sq cm (31 sq in)

A popular myth

It's a popular myth that it takes more energy to produce hydrogen (by separating it from other elements in molecules) than hydrogen provides when it is converted to useful energy.

A little energy input will, in fact, release the explosive energy stored in the hydrogen bonds themselves, as this research paper shows experimental results from the kinetic energy of cold fog, generated in a water arc plasma:

*"With 50 J of input energy, the quantity of fog produced is of the order of 0.75 g of water. To dissociate this amount of water into oxygen and hydrogen would require 10 kJ of energy. Hence the fog explosion is unlikely to be caused by electrolytic dissociation of water molecules. Without this dissociation, the most likely source of the explosion energy is that **stored by hydrogen bonds** between the water molecules. This bond energy is said to be equal to the latent heat of evaporation, and therefore could contribute up to 2200 J g–".*

From the scientific research paper
"Arc-liberated chemical energy exceeds electrical input energy[1]"

PETER GRANEAU, NEAL GRANEAU,
GEORGE HATHAWAY and RICHARD L. HULL
Centre for Electromagnetics Research, Northeastern University, Boston, MA 02115, USA
Department of Engineering Science, University of Oxford, Oxford OX1 3PJ, UK
Hathaway Consulting Services, 39 Kendal Avenue, Toronto, Ontario, Canada M5R 1L5
Tesla Coil Builders of Richmond, 7103 Hermitage Road, Richmond, VA 23228, USA

(Received 11 December 1998 and in revised form 5 August 1999)

Arc-liberated chemical energy exceeds electrical input energy:
https://bit.ly/3KRaXhH

Electrolysis Equation

Water + Energy   Hyd + Oxy + Heat

$$2H_2O + 4e^- \rightarrow 2H_2 + O_2 + Heat$$

Or to put it another way...

This equation may look equal, but we just released a sleeping dragon.
Water on the left and rocket fuel on the right!

In other words, water contains hydrogen and oxygen and with a spark of energy, it can release a massive amount of energy and heat. NASA's Space Shuttle, for example, carried tanks of Liquid Hydrogen and Liquid Oxygen as a propellant. The Oxygen acts as an oxidiser as there isn't any in space.

Another example

The hydrogen-filled Hindenburg notoriously exploded as it was coming in to land, in New Jersey, United States, on May 6, 1937, probably due to the electrical storm causing a lightning strike. The rain-sodden wet guy ropes would have made a perfect path to earth, completing the circuit. Lightning is an extremely high voltage with a relatively low current. (No energy would have been 'added' to the hydrogen as far as I'm aware, as it was only used for its lighter-than-air properties! and yet, we have a massive explosion).

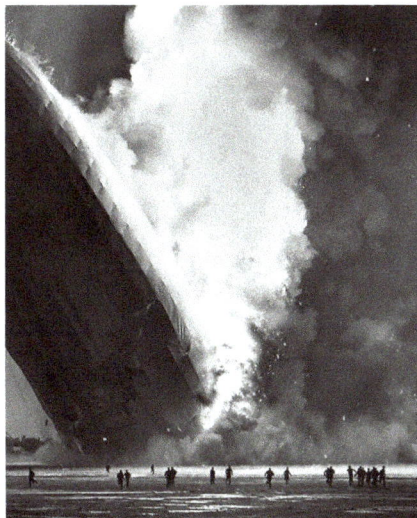

99

Michael Faraday

When Michael Faraday (a bookbinder and hobbyist) published his 'Laws of Electrolysis' in 1833, it was the middle of the steam age, when the best technology for vehicle locomotion was to burn lumps of coal to boil water, using the resulting steam pressure to drive the wheels. The internal combustion engine was only 35 years old at that time. (To give it some context, Queen Victoria was just a 14-year-old princess and William IV was the king of England). Since then, many advances have been made in science and technology. Many many refinements have been made to engines of all types and propulsion systems for everything from motorcycles to space rockets. Anyone who quotes the Laws of Electrolysis as an example of how using water as a source of hydrogen can't be done efficiently, is simply stuck there in the middle of the 19th Century. For a start, any understanding of what is happening at a molecular level wasn't possible at that time, as the electron hadn't even been discovered (it was discovered 64 years later).

Faraday's primitive equipment

All this nonsense that you often hear about 'putting energy in' to get equal energy out is based on the inefficiencies of Faraday's primitive 19th-century apparatus. Faraday used it to formulate his laws of electrolysis, including the law that the same quantity of electricity, no matter what its source, decomposes the same quantity of matter. However, he was not using the most efficient method. Faraday was only using a DC (direct current) electricity supply for his famous electrolysis experiment along with tiny pieces of wire for the electrodes.

"I made such simple experiments in chemistry as could be defrayed in their expense by a few pence per week, and also constructed an electrical machine, first with a glass phial, and afterwards with a real cylinder as well as other electrical apparatus of a corresponding kind."

Michael Faraday

The Quest for Efficiency

A major improvement to the efficiency of electrolysis is the use of an *alternating current*. If a *resonant frequency* is used, it produces much more hydrogen with a lot less energy input. I recently performed the standard electrolysis experiment as taught in schools and colleges around the world. It took about an hour to produce a quarter litre of hydrogen by splitting water with a 9v DC supply. The tiny piece of platinum foil struggled to fill an upturned test tube. Modern equipment, using a pulse width modulator and a dry cell can achieve about four to six litres per minute, that's eight to twelve times more efficient! Resonant frequency electrolysis doesn't break the Laws of Electrolysis, it re-defines them, simply because it's more efficient.

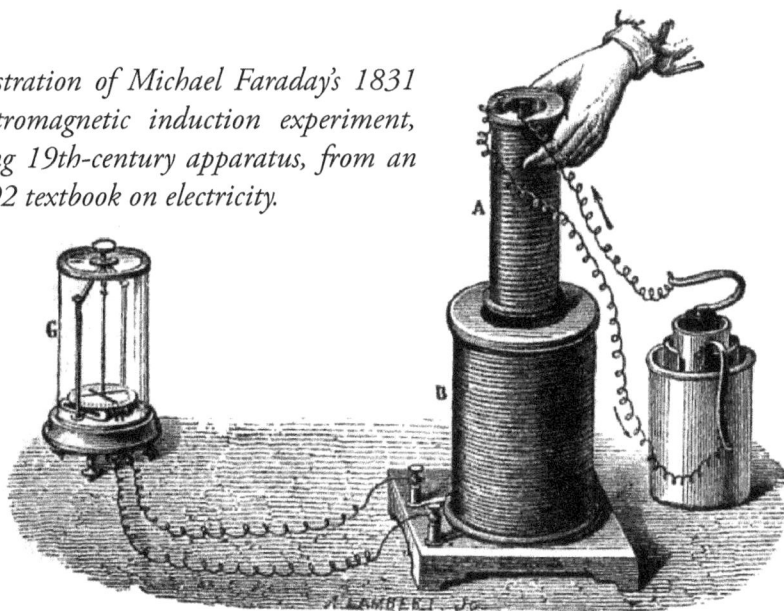

Illustration of Michael Faraday's 1831 electromagnetic induction experiment, using 19th-century apparatus, from an 1892 textbook on electricity.

Another thing that makes electrolysis efficient is plate size. Modern multi-plate fuel cells have vastly larger gas-producing areas and are therefore far more effective. Bob Boyce, for example, used 101 plates of approximately 6" (152.4 mm) square (each with two effective sides) in his drycell. That's a total area of 1,212 square inches (7819 sq cm), a huge difference from using a tiny piece of platinum wire and all with the same voltage running through (amounting to about 2v per cell between plates). Remember how in the chapter on The Power of Water ('A lesson in efficiency from Stephenson's Rocket Locomotive') we talked about his innovation of multiple tubes massively increasing the area of contact with the steam, which allowed the increased efficiency to make the process viable?

There is in fact far more energy already contained within water than Faraday could possibly have known. (In fact, there is three times as much energy in water than in the equivalent weight of fossil fuel).

When talking of fossil fuels people usually count the energy within it (e.g. Octane rating), but never mention the energy that it takes to extract or refine it. Conversely, with water as fuel (using electrolysis) they talk about how much energy needs to be 'put into it', but never mention the highly combustible combination of hydrogen and oxygen that's already locked within it. In other words, just like all fossil fuels, it possesses chemical energy! Hydrogen is common to both fossil fuels and water, but water contains the ideal oxidant for combustion, which is oxygen, rather than the inferior carbon, (which combusts very poorly, incompletely and therefore very inefficiently, especially when combined with only 21% oxygen in the air). Not only that but also, when anyone says that the process of fracturing water to release hydrogen and oxygen is inefficient, the quoted figures are based on the best results of Faraday, using primitive equipment and the scientific knowledge of almost 200 years ago.

The electrolysis method they teach in school is a 'brute force' process where a single direct current voltage is used to create a few tiny bubbles of hydrogen and oxygen. The method proposed in this book, namely the use of a resonant frequency coupled with a gated pulse, uses an alternating current at a certain frequency that matches that of the water. The induced fluctuating magnetic field flexes the water molecules at about 500 times per second until they break, separating the hydrogen and oxygen atoms. Now, the energy released from the water far exceeds that of the low alternating current input.

Faraday used a simple direct current which of course has zero frequency. That's the difference between pushing something and hammering it. Imagine the energy required to push a nail into a piece of wood by hand? In fact, it's probably impossible. However, if you take repetitive swings at it with a hammer, most people would be able to do it easily. The reason is that it's swinging the weight of the hammer many times that works best. The larger the swings and the more frequent the blows the better. You don't need to push at all. Continually pounding something at its resonant frequency is even more powerful. The fact that an opera singer can break a glass, just by singing a note at the correct frequency and sustaining such a note, shows that minimal energy is required. The glass, subjected to the vibrations undergoes catastrophic deformation. Similarly, continual flexing 500-600 times a second breaks the molecular bonds of water and releases the gases.

What are the Laws of Thermodynamics?

Whenever anyone talks about using water as a fuel source, people often quote the Laws of Thermodynamics, saying that these laws are violated and therefore the process is not even worth looking at. Thermodynamics is a branch of physics that deals with heat and other kinds of energy. It specifically deals with the conversion of one form of energy to another. You can see how this could be very useful in disciplines such as meteorology as well as engineering and many other areas of science.

There are basically four laws:

- The First Law is called the law of conservation of energy, in which energy can be transformed, but it can't be created or destroyed.

- The Second Law defines a closed system in which energy stays constant or increases due to outside forces.

- The third Law is the law of "absolute zero," that is the extremely low temperature of –273.16°C (–459.69°F), in fact, the coldest temperature possible where no heat can exist at all. Which is not believed possible.

- The fourth law, (oddly called the 'Zeroth Law'), states that if two thermodynamic systems are in thermal balance with a third system, then they are also in thermal balance with each other.

Using modern advanced techniques, including electronics, acoustics and a real understanding of molecular chemistry, we can achieve results in excess of 20-30 times the efficiencies of Faraday's time. There's a massive amount of energy locked in water, that we can now access easily using modern techniques. And yet, people still insist on exclaiming "the energy you have to 'put in' equals the energy that you get out". Not only is that an absolute myth, but it's a barrier to the progress of this important solution to the world's energy crisis.

No laws are violated

Firstly, with electrolysis, we have an open system in which there is often energy involved which is not taken into account. For example, static and magnetic energy and the energy already contained within the elements. We are adding energy in the form of electricity and the reaction takes place in a vessel open to the air. The resulting gases, hydrogen and oxygen, escape into the air and can be collected, together, or separately. We are not just adding energy as electricity, *there is also chemical energy within the hydrogen and oxygen molecules*, in the form of powerful attractive forces between the hydrogen and oxygen bonds within water. Why is this always overlooked? It's much more combustible than fossil fuels.

With modern material science, coupled with scientific knowledge and advances in precision technology, especially electronics (the science of moving electrons), we can achieve amazing things.

Faraday would surely be blown away by modern technology and yet some people mistakenly cling on to his outdated calculations as if they were set in concrete.

The 'theory' of electrolysis

Yes, you read that correctly. Don't get me wrong, Faraday was a great man and we owe him a debt of gratitude for his discoveries, such as the electric motor and generator, but he should never have called his conclusions on electrolysis a 'Law'.

The point worth making here is that scientific 'Laws' are not laws like those of our legal system. Laws in science are often a distillation of one, or many scientists' observations using state-of-the-art equipment and interpretation based on their current knowledge, which of course changes over time. For this reason, laws can be disproved and modified to reflect the latest understanding.

We may not have come across a unique set of circumstances that lead to a particular phenomenon occurring, or, at least, it may not have been observed. A theory can be updated, added to, or even scrapped completely in the light of new evidence, but a 'law' is fixed until perhaps it is 'repealed'. Yes, you can revoke a law, but by its very nature, that's quite a drastic proposition. So, a 'theory' naturally lends itself to being updated. It welcomes new evidence.

Science doesn't have all the answers

Gravity is a force that we don't fully understand but is used for its power. (For example, Hydroelectric dams generate electricity through the weight of water driving giant turbines).

The electrical potential of the earth's atmosphere, which causes massive electrical storms and lightning sparks goes into the millions of volts. Benjamin Franklin (Philadelphia USA 1706 - 1790) experimented with a kite held on a string in a thundercloud, to prove that it was indeed electricity and as a result invented the lightning rod seen on many buildings, intended to 'earth' the electric potential harmlessly away. Tesla also experimented with the earth's electric potential and famously had plans to harness it on a global scale, but never realised his dream.

(Negative electricity is an unfortunate choice of terminology as it doesn't mean a lack of electric charge, it simply means that it is flowing in the opposite direction. Plus, it depends where we are measuring it from. Electric potential is always the difference between two measured points).

Static electricity means a potential that is literally 'not moving', but when it does, it can be millions of volts, such as lightning or the sort of shock you get from your car door handles. It is not usually harnessed in any useful way. In electronic terms, it is sometimes called 'back EMF' (electromotive force) or Counter-electromotive force (CEMF) and is usually considered a nuisance, as it opposes the current flow that we are trying to achieve, but it can in fact be harnessed. (Many of the so-called 'free energy' devices use this principle). Whenever there's an electrical current there's always a surrounding magnetic field. It's an interplay between the two. Magnetism is also a natural force that science cannot fully explain.

Moving magnets can create electric currents and electric currents can move magnets creating useful force in motors and solenoids which can be found everywhere in our modern world. These often-overlooked forces can resist or enhance other uses of electrical power. Let me give you an example - Regenerative braking. That is when in an electric-powered vehicle, the resistance of an electric motor is used to slow it down and bring it to a stop, generating energy back into the batteries. Often energy is lost as heat and engine cooling usually results in greater engine efficiency. Turbocharging is a clever way of enhancing engine performance, by returning what would otherwise be wasted hot exhaust energy to boost the power.

So what's my point? The technology of water-splitting is all about efficiency. Unless you take into account all the forces involved, including temperature, pressure, magnetic charge, electrostatic charge, chemical energy, atmospheric energy as well as the energy that you are putting into the system, then you will have little chance of success.

Saying that you can only get the same energy out as you put in, is a very simplistic way of looking at it. Nobody seems to question the energy contained within the carbon/hydrogen content of fossil fuels or the considerable amount of energy that it takes to refine crude oil into a usable form. (It's distilled up to 400 °C).

Overunity? Perpetual motion?

Sometimes people talk of 'over-unity' which is also nonsense, as the water, which contains energy, is being turned into usable gas and therefore diminishes with time in exactly the same way as any type of fossil fuel. Another word for 'over-unity' is a 'perpetual motion system'. Calling it that is just another way to discredit the science behind it. Using an efficient way to split water molecules to extract the energy contained within, is by no means 'perpetual motion'.

Free energy

The term 'free energy' is also a bit of a misnomer and is often assumed to mean energy that has no cost in energy terms. For that reason free energy devices are often classed as pseudoscience. However, 'free energy' can also mean energy that is freely available from the environment. (A perfect example is perhaps the Bedini motor[1], which is in fact using the normally wasted back e.m.f.).

1. Bedini motor patent: https://patents.google.com/patent/US6545444B2/en

The Bedini Motor

The Bedini Motor is often touted as a 'radiant energy battery charger', but the pulses generated are ideal for water-fuel electrolysis. It's a monopole electromagnetic motor which has proved to be far more efficient than most other designs currently in use. I'm not going to go into detail here, but if you do look at the design you will see many parallels to much of the technology that I've covered so far (using coils, capacitors, electromagnets and inductors to create a resonant output). The main principle is that it utilises the 'flyback EMF', a high voltage spike produced by a collapsing magnetic field that normally causes drag on a motor and is therefore counter-productive. For some inexplicable reason, devices like this are often considered to be pseudoscience.

The patented Bedini SSG (Simple School Girl's) Motor, by John C. Bedini, so-called, as it's so simple a 9-year-old girl could make it (after a nine years old girl did just that and won a nodal science award for doing so).

There's also a very similar, incredibly efficient device by Robert Adams of New Zealand, former Chairman of the Institute of Electrical & Electronics Engineers, Inc., U.S.A., (N.Z. Section)., the 'Pulsed Electric Motor Generator[1]' that's worth a look. Again the authorities tried and to some extent succeeded in suppressing it. I wonder why?

1. Pulsed Electric Motor Generator: https://bit.ly/38ZCCQN
2. The 800% efficient Robert Adams motor-generator:
   https://www.youtube.com/watch?v=J2bPDDWqSvM

Using AC for electrolysis

Why does AC electrolysis break the hydrogen/oxygen bonds more efficiently?

As an analogy, if you want to break a strip of metal, you could hit it really hard in the centre whilst supporting each end on something solid. Otherwise, you could gently, but firmly, bend it back and forth until it breaks through metal fatigue. That repetitive motion is like the alternating current switching back and forth. What is actually happening is that during electrolysis you are subjecting the molecules to an alternating magnetic field and then an electric field. The collapsing fields apply a destructive force on the water molecule, which is, in effect, a small magnet in itself. (The hydrogen atoms are negatively charged and the oxygen molecules positively charged).

Using resonance is far more efficient because it is voltage and frequency that does the work, not current.

Faraday didn't use a frequency at all, just a DC current. No wonder he believed that you only get as much energy from the water as you put into the reaction by passing a low current. If he had released 8 litres per minute of gas, as is now possible with a resonant electromagnetic field, he would have surely come to some considerably different conclusions. Also, he wasn't factoring in the huge amount of chemical energy that water possesses. You don't need to 'add energy' at all, by just using enough applied in the correct way to release the massive amount of chemical energy that is already within it. After all, how would he know that you could unleash rocket fuel mixed with Hindenburg - destroying power?

Hydrogen combined with oxygen is even more powerful than the separate gases, as oxygen aids combustion, the two releasing massive amounts of energy as they implode by a factor of 1860 times as they recombine into water. Also, trying to say that high-frequency, resonant electrolysis breaks the laws of thermodynamics is simply laughable. Bear in mind though, that it's usually those people whose livelihood or finances would be destroyed if we stopped using fossil fuel overnight.

The carbon that comes with hydrocarbon fuels is just a nuisance. It doesn't burn well, clogs up fine machinery and fills the atmosphere with unburned noxious gases and unhealthy microscopic particulates. Water, however, comes instead with a good proportion of oxygen, which as any petrolhead will tell you, massively aids combustion. So we have the ideal combination. Hydrogen has a very pure flame, which burns easily and instantly. Making it on-demand from water means you overcome the problems of hydrogen storage. Any kind of water can be used and the technology can even be used for water desalination and purification. Perhaps it's time to start looking seriously at this technology and stop silencing, imprisoning, or even murdering its innovators (e.g. The likes of Andrija Puharić, Stanley Meyer and Bob Boyce in later chapters), Suppressing planet-saving technology like this in the name of profit is incredibly stupid.

I realise that I'm labouring my point, but the key to my whole argument here is that there is a way of doing electrolysis that is far more efficient at releasing the 'already volatile' hydrogen from the water. To put it another way, clean hydrogen plus oxygen from water is far superior to filthy carbon with hydrogen from fossil fuels as a means of powering our world. The by-product, water. Pure, clean, green energy, no pollutants, no carbon deposits, no carbon footprint, no climate change. Result!

The oil companies have had their day and the damage being done is becoming irreparable.

So please, it's time to take this technology seriously.

So electrolysis is at the heart of water-splitting to release hydrogen and oxygen, as we can't simply burn water as a fuel. But as we have seen, conventional DC electrolysis isn't efficient enough to be useful. Instead we can use an oscillating resonant signal to shake the water molecules apart. Notwithstanding that, even a larger plate size for the anode and cathode, linked to an electronic pulse can also work wonders. Let's look at the HHO movement.

9

HHO

Also known as 'Brown's Gas' Hydroxy or Oxyhydrogen

The mysterious explosive combination of Hydrogen and Oxygen

If you do a search on YouTube, eBay or Amazon, you'll doubtless find many examples of people experimenting with 'HHO'. This means a combination of hydrogen and oxygen gas which needs to be made on demand and used immediately as it naturally wants to quickly recombine back to water. They perhaps offer kits that split water to produce HHO as a way of adding a hydrogen/oxygen mix to a standard engine in order to boost economy. I myself had the carbon cleaned from my engine using such a system.

HHO or 'Brown's Gas'.

'HHO', 'Oxyhydrogen', hydroxy, or 'Browns Gas' after its discoverer, Bulgarian Professor Yull Brown[1] (1922 - 1998). It's even more explosive than hydrogen and oxygen separately. So why have you probably not heard about it? It's a combination of hydrogen and oxygen gas, which is unstable as it wants to instantly recombine back to water, so it has to be created on-demand and used immediately. I guess something so fleeting and impermanent didn't seem worthy of study. However, it burns far more efficiently than hydrogen alone as it contains the right proportion of oxygen for perfect combustion, which is a big deal when you consider that the atmosphere only contains about 21% oxygen. Some people claim that if you add HHO to an engine, you can achieve a 35% fuel saving. Imagine what could be achieved if the research and development teams of major motor manufacturers started to study this. And this isn't even using the higher efficiency 'resonant frequency' principles outlined previously in this book, although a 'PWM' device does pulse the electrical input.

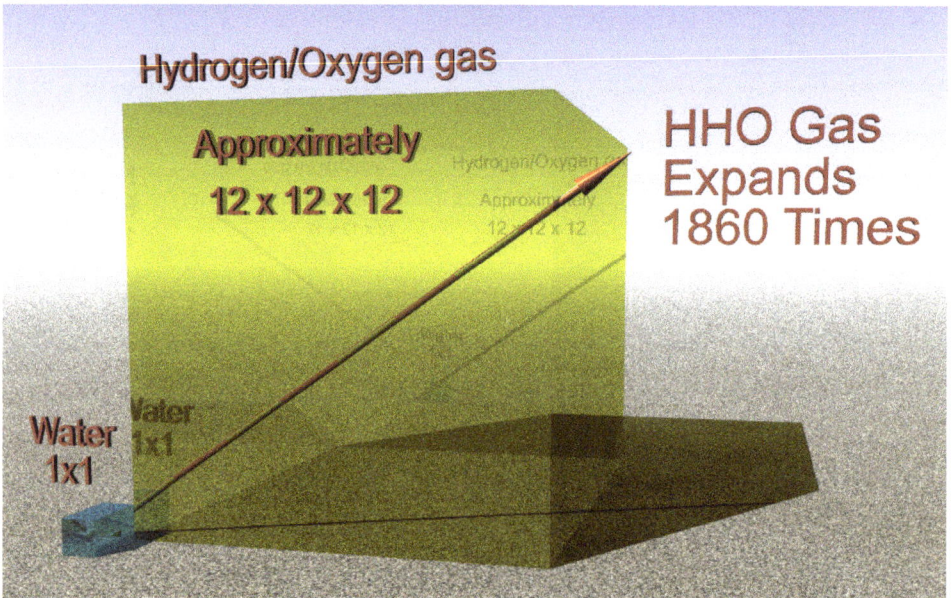

HHO gas, a combination of the two Hydrogen and Oxygen gases released from the water, expands to 1860 times its original volume, under normal temperature and pressure. (e.g. That's slightly more than a 12 x 12 x 12 cube from a 1 x 1 cube). Conversely, if ignited, it will implode to the same extent.

1. Yull Brown (aka 'Ilia Valkov see "HHO Generators – Original Brown's Gas Company" https://yullbrownsgas.com)

Improving your vehicle's mileage using HHO

Do a YouTube search on the term 'HHO' and you will find many videos on the subject. It seems that many people are keen to increase their fuel mileage by injecting this combination of oxygen and hydrogen split on-demand from water, into the air intake of their vehicles. It's possible to make savings on fuel costs of between 10% and 50% with most of these systems. In fact, there are now many individuals and companies offering full kits to make the process easier. (However, the kits are fairly simple to fabricate yourself from a few simple, easily obtained parts). It seems that people are keen to improve the mileage they are getting with regular fuel rather than replacing the use of fossil fuels. I believe that this is because adding HHO is relatively easy. Running a vehicle entirely on hydrogen and oxygen sourced from water is a much more challenging proposition. Rather than use a 'water fuel cell' similar to the designs of Stanley Meyer, these systems use a form of dry cell as seen in the previous chapter. As the water is contained within, they are far more suitable for vehicle use than the type that has a set of tubes sitting in a flask of water. Adding hydrogen to a fossil fuel-powered internal combustion engine is one thing, but if you are going to run an engine entirely on hydrogen, or a combination of hydrogen and oxygen, then there are a few important considerations.

Electronics for HHO

Most modern vehicles employ sensors which feed back to a central computer, called an EMU', or 'Engine Management Unit'. Or 'Electronic Control Unit' (ECU). One of these is an exhaust gas temperature sensor (EGT), which tells the computer when the exhaust is getting too hot and has the effect of adjusting the amount of fuel being injected. A very lean engine will run hot, but when adding HHO we need to fool the vehicle's computer as we are adding fuel outside the manufacturers' usual configuration. This also applies to another sensor called the 'EFIE' (Electronic Fuel Injection Enhancer). Surprisingly, these systems don't ever seem to use a resonant frequency. I've not seen any evidence of anyone using a resonant frequency with a dry cell, although many employ a 'PWM' device, or 'Pulse Width Modulator', which pulses the power on and off rapidly. This is mainly to prevent 'thermal runaway', which occurs due to the hydrogen flame burning really hot and also tending to draw more and more amps (power) through the dry cell which also heats up. This PWM device can be adjusted to prevent that manually or automatically.

HHO kits for vehicles

A typical HHO Kit for vehicles usually consists of a multi-plate drycell, a water reservoir and a PWM driver circuit. This add-on is designed to inject the hydrogen/oxygen combination directly into the engine's air intake (preferably after the Mass Airflow Sensor), in addition to the existing fuel. It also includes a bubbler to prevent any flame from getting back to the drycell and igniting any accumulated gas, although that will be minimal.

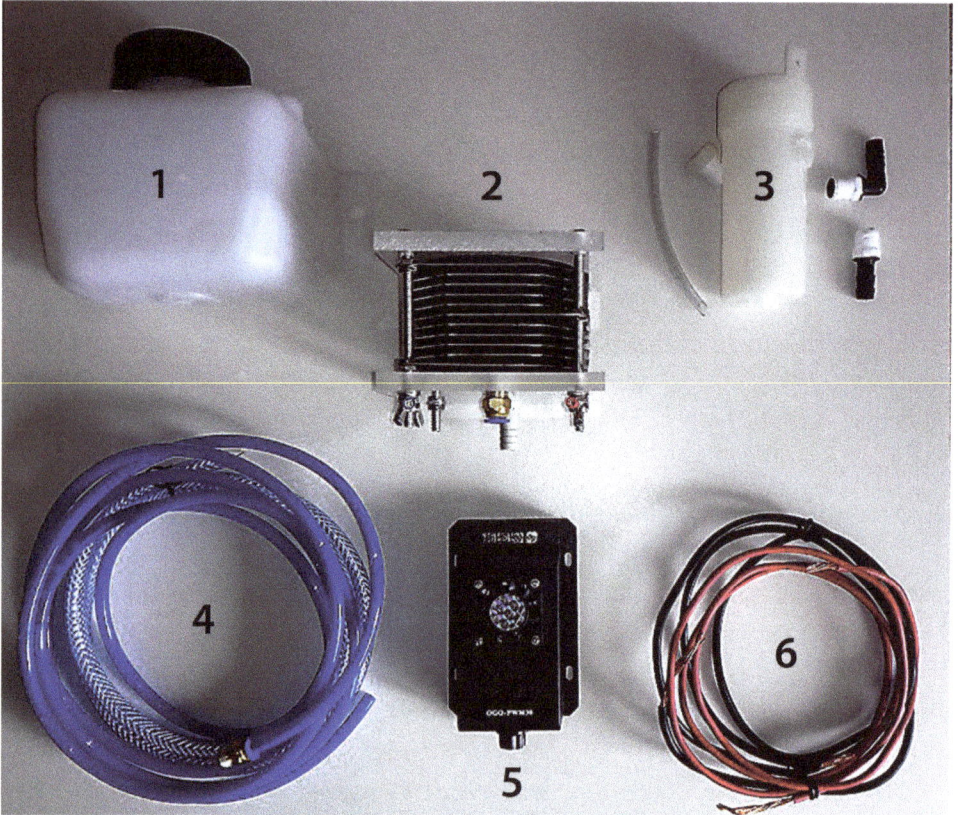

A typical HHO Kit for a vehicle. (Photo by the Author):
1. Water reservoir, 2. Drycell, 3. Bubbler, 4. Tubing (for water and gas),
5. PWM device, 6. Wiring.

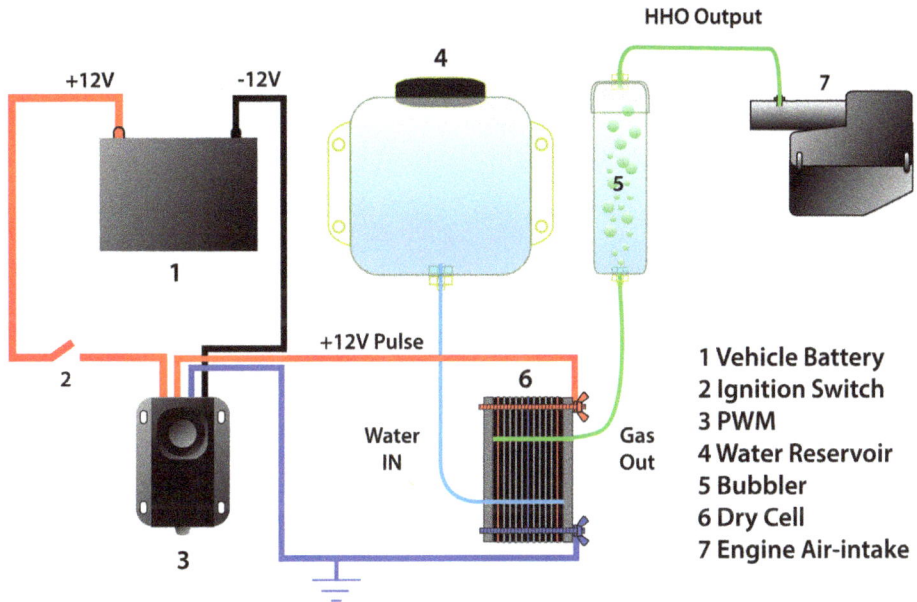

Schematic by the author, showing connections between components of a typical HHO Kit for a vehicle.

What is 'PWM' Pulse Width Modulation?

PWM Pulse Width or pulse DURATION Modulation (PDM), is a way of modulating the amount of power delivered to a device by very rapid switching of the current. It is used to control the current applied to a motor, for example, in order to finely control the speed, but what we are interested in here is the control of the output of hydrogen in an HHO setup. It's not exactly a 'resonant frequency' but it is a different kind of waveform, which also has the effect of pulsing the current. Your heartbeat creates a pulse as it pumps your blood around. Anything that works in cycles creates a 'waveform' like the heartbeat trace on a hospital monitor. A PWM circuit also delivers a 'pulse'. Essentially it is a 'square wave', as it switches the current 'on' for a short period, then 'off', pauses and repeats. This is called the 'duty cycle'. The duty cycle is expressed in percentages. (If it is 'on' constantly that is a duty cycle of 100%, but if it is only on for half the time that represents a duty cycle of 50% and effectively delivers half the current). The period of time it is 'on', 'off' and the pause between is easily controlled by an electronic oscillator, as we saw in an earlier chapter.

Now, research into HHO gas is relatively new. Perhaps because it immediately wants to recombine back to water, so is really only of use if you are going to create it on-demand. It cannot, therefore, be stored anyway, unless it is separated into oxygen and hydrogen. However, as burning fuel is far more efficient with a good flow of oxygen, hydroxy gas is ideal for combustion in an engine as it already carries enough oxygen along with it, possibly in ideal proportions. I believe from my observations that it has some special properties that need further investigation.

So, using HHO gas in an engine will help to burn the fuel more efficiently and completely, cutting down on emissions. Since emissions from heavy industry account for about 22% of global CO_2 emissions (greenhouse gas), HHO gas seems to be an ideal solution, indeed with additional benefits.

Many people claim improved fuel consumption as a result of using these kits when used in combination with existing regular carbon-based fuel and even on large heavy goods vehicles with much bigger engines.

An excellent article on the effects of adding HHO to an internal combustion engine, by Science Direct can be found here: https://www.sciencedirect.com/science/article/pii/S1110016815001714

'Hydroxy Gas' (trademarked)

Hydrogen and Oxygen combined, as they will be if not separated by the electrolysis process, is an unstable gas and will return to water unless immediately ignited. Also when under combustion will actually *implode* becoming water once again. Any petrolhead will tell you that getting more oxygen into an engine increases the power output. However, there is only about 21% oxygen in the atmosphere which also includes nitrogen and other gases. Hydroxy already includes 33% oxygen to 66% hydrogen, making it the ideal combination for efficient combustion.

There is a demonstration of its effectiveness in a YouTube demonstration video showing three balloons being ignited[1]. One filled with just hydrogen that goes off with a loud 'pop' when burst using a flame torch, one with just oxygen that bursts undramatically, just like a balloon filled with only air. But when the balloon filled with HHO gas bursts it really explodes dramatically.

1. Three Balloons Demonstration: https://www.youtube.com/watch?v=a6qGIMqDKwA

Hydrogen the carbon destroyer!

Carbon deposits build up within an engine over a period of time, causing a huge loss of efficiency along with increased wear and tear. If you inject hydroxy gas directly into an engine through the air intake, it has the effect of not only improving combustion but also cleaning carbon out of the system, the overall effect being improved fuel economy and better performance.

Smoke test

I had a 15-year-old camper van with an 18-year-old engine that failed its MOT test (the UK's Ministry of Transport test) on its poor 'smoke test' emission results. I took it to a 'Carbon cleaning' specialist who connected it up to a portable electrolyser that splits water into hydrogen and oxygen and injects it into the air intake for about half an hour. While running the engine, the improved combustion cleaned out all the carbon from the engine including the catalytic converter. The result was that it passed the test with flying colours. The official readings were 3.21 before the clean-out and 1.21 afterwards (see the scans on the next page).

Cleaning carbon from a high mileage engine using HHO gas. (Photo by the Author).

FAIL!
Clogged up with carbon deposits.

MOT (Ministry of Transport) Emission Test FAIL before introducing HHO to the engine.

VTS Name:	WINDMILL GARAGE
VTS Address:	
VTS No.:	82004

Date & Time: 9/9/2021 15:13		**Tester ID:**	
MOT No.:		**VRM:**	
Make:	FORD	**Model:**	TRANSIT
VIN:	K19567	**Size (cc):**	2402

Crypton - Diesel MOT Smoke Test program	Program - Version 2.1.0.20949

MOT SMOKE TEST - Turbo
Manufacturer's plate illegible

RESULT				DIAGNOSIS	LIMITS	
					Min	Max
Oil temperature	=	83	°C	-	60	-
Smoke Reading						
Peak 1	=	4.19	m^{-1}		-	-
Peak 2	=	4.03	m^{-1}		-	-
Peak 3	=	3.10	m^{-1}		-	-
Peak 4	=	4.01	m^{-1}		-	-
Peak 5	=	3.10	m^{-1}		-	-
Peak 6	=	2.71	m^{-1}		-	-
Zero Drift	=	0.06	m^{-1}	**PASS**	-	0.16
Average	=	3.21	m^{-1}			
MOT Test Result						
Turbo	=	3.21	m^{-1}	**FAIL**	-	3.00

Crypton Smoke Analysis
© Crypton 2003 - 2018

118

PASS!

4 Days later
Just look at the reduction in smoke readings

MOT (Ministry of Transport) Emission Test PASS after introducing HHO to the engine air-intake for 30 minutes. The Hydrogen/Oxygen improved combustion has cleaned the system of carbon deposits and particulates from the engine and exhaust system.

VTS Name:	WINDMILL GARAGE
VTS Address:	
VTS No.:	82004

Date & Time: 13/9/2021 15:26		**Tester ID:**	
MOT No.:		**VRM:**	
Make:	FORD	**Model:**	TRANSIT
VIN:	K19567	**Size (cc):**	2402

Crypton - Diesel MOT Smoke Test program	Program - Version 2.1.0.20949

MOT SMOKE TEST - Fast Pass
No Manufacturer's plate visible

RESULT				DIAGNOSIS	LIMITS	
					Min	Max
Oil temperature	=	86	°C	-	60	-
Smoke Reading						
Peak 1	=	1.22	m⁻¹		-	-
Zero Drift	=	0.01	m⁻¹	**PASS**	-	0.10
Average	=	1.21	m⁻¹			
MOT Test Result						
Fast Pass	=	1.21	m⁻¹	**PASS**	-	1.50

CRYPTON	Crypton Smoke Analysis © Crypton 2003 - 2018	

Catalytic (flame-less) heating

It's quite easy to make an effective catalytic (flame-less) heater using a vertical metal tube full of fine sand or other non-combustible materials. FreeFromFuel.com has full instructions for how to do this using a vertical tube of any kind of metal (perhaps iron, steel, copper, or aluminium), with HHO dry cells providing the gas. There's also a YouTube video demonstrating it[1]. Imagine what savings can be achieved if we used this for heating. This has massive benefits over conventional gas or paraffin heaters, which emit poisonous fumes, including carbon monoxide.

In order to work, the sand needs to be preheated to about 60°C to kick start the catalytic reaction. This setup will only require between 0.5 & 3 litres per minute of gas. Any more hydrogen gas will escape from the top, so needs to be adjusted to prevent that. There may be some condensation as the oxyhydrogen converts back to water which will need to be drained off. The addition of a layer of platinated sand produces even more heat. (Full instructions in the excellent publication FreeFromFuel[1], in German or English, which also has plans and instructions for building complete HHO systems, including dry cells).

The perfect cutting tool

HHO can also be used to power a torch flame, similar to oxy-acetylene, but burns with an extremely pure flame. In fact, hydrogen burns with an invisible flame and will melt almost anything including steel and rock. The flame temperature is only about 140°C, about ¼ of the heat of a candle, but when directed at anything solid increases to around 2,800 °C (5,100 °F), hotter than a hydrogen flame in air. So it seems to vary depending on what it is reacting with. What is happening is that the oxygen and hydrogen recombine as an implosion (1860 litres of gas > 1 litre of water) and release a huge amount of heat. That's fusion! It's even possible to burn this flame underwater although it doesn't heat the water up! It will still melt rocks under the water although it doesn't react with the water or the atmosphere.

1. HHO Workshop "Free From Fuel": Heating With Oxyhydrogen (DVD Chapter 2): https://www.youtube.com/watch?v=b7UPMSleuek
2. Book and videos on DVD about constructing HHO Systems: www.FreeFromFuel.com

Batteries vs Drycells

There are obvious similarities between batteries and drycells as they are both made of alternating metal plates immersed in a liquid catalyst. Batteries also release gas when they are charging. There is an intriguing patent for an electrolyser that is in fact a combination of the two[1].

Seawater hydrogen generator

Bolivian Francisco Pacheco's invention needs no external power supply as it produces its own by virtue of its electrochemical makeup. Alternating plates of magnesium and aluminium create an electric charge and the seawater catalyst is then split into hydrogen and oxygen. As the catalyst contains salt, there is also chlorine gas produced but found to be negligible. The inventor tried to publicise his invention without much success, despite powering a boat for several hours with seawater. I am totally dumbfounded as to why that would be. We will look at his invention a little more in chapter 17.

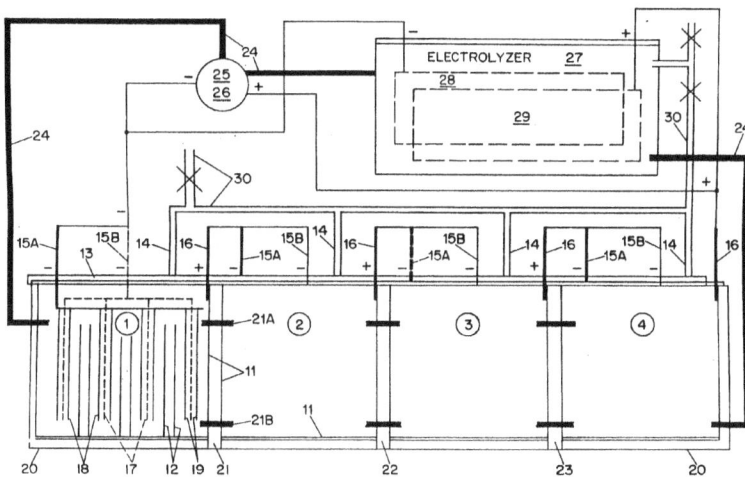

FIG. 2

The Francisco Pacheco Hydrogen Generator

1. The Francisco Pacheco Bi-polar auto electrolytic hydrogen generator:
    https://patents.google.com/patent/US5089107A/en?oq=5089107

So, many people seem to be getting good results from HHO gas released from water, despite the fact that I've seen no evidence of any of them using a resonant frequency to improve efficiency. Using a gated pulse is however a form of frequency that seems to work to some extent. Let's go on to look at the main pioneers of the resonant frequency method for water splitting.

10

Henry (Andrija) Karel Puharić

(1918 – 1995) (a.k.a Henry)
The pioneer who discovered the use of resonant-frequencies to split water molecules

Andrija Puharić (by kind permission of his son Andy Puharić)

Born of Serbian immigrants, Puharić lived in the USA.
As a medical specialist, he realised that sound waves can break up dangerous blood clots that cause deadly thrombosis. Furthermore, he used the exact same technology to split the water molecule which he meticulously observed at a microscopic level. His story is rather technical, but a fascinating one and key to understanding the theme of this book: Resonant frequency water splitting.

123

"Disclosed herein is a new and improved thermodynamic device to produce hydrogen gas and oxygen gas from ordinary water molecules or from seawater at normal temperatures and pressure...

...The hydrogen, in gas form, **may then be used as fuel**; and oxygen, in gas form is used as oxidant. For example, the thermodynamic device of the present invention may be used as **a hydrogen fuel source for any existing heat engine - such as, internal combustion engines of all types, turbines, fuel cell, space heaters, water heaters, heat exchange systems, and other such devices. It can also be used for the de-salinization of sea water, and other water purification purposes**. It can also be applied to the development of new closed cycle heat engines where water goes in as fuel, and water comes out as a clean exhaust".

Andrija Puharić From Patent US 4,394,230 1983-07-19

Henry Puharić, born Andrija Karel Puharić (1918 – 1995) (a.k.a 'Henry' which was his parent's Anglicised version of 'Andrija'), was a Doctor of Medicine, a medical researcher and inventor who was born in Chicago, Illinois, and was one of seven children born to Croatian immigrants. His father had emigrated from what was then the Austro-Hungarian Empire, entering the U.S. in 1912 as a stowaway. He later became a lecturer at Northwestern University Medical school, in Evanston, Illinois.

Patents

Puharić brought out many patents covering a variety of subject areas, notably medical applications, particularly as an audio specialist. Two of his 50-plus patents were devices that improve hearing. Of particular note are his research into audiology and the effects of electrical stimulus on the human body and mind. It was only a small sidestep to studying the effects of an electrical frequency on water molecules. Through his work as a medical scientist, he had already been working with magnetic fields and electrical stimuli to manipulate another liquid, blood, within the veins. Particularly true in the case of, for example, thrombosis, or blood clots, which can be deadly and certainly cause a lot of harm if they travel to the brain. He had found a way of using sound waves with a particular *resonant frequency* to break up the clots and render them harmless. Also, other patents that he submitted, proposed using similar technologies to deal with hearing problems. (This is so closely related that it shares the same electronic circuits to create the waveform). Perhaps medical science should take a look at these other patents which contain his solutions, but that's another subject although related in many ways.

He did a great deal of in-depth research into splitting the water molecule into hydrogen and oxygen for use as a fuel. His hydrogen sourced from water used a similar principle to break down the water molecule using an audio tone generator and amplifier to create a 'resonant frequency'. He claimed that he did many hours of research to get to that stage and had to be careful of the cost of testing with the equipment. (Despite being an acquaintance with the billionaire David Rockefeller!).

Rockefeller, the oil baron, who instead of helping Puharić, slapped a gagging order on him.

Of his 50-plus patents, the one we are most interested in here is the U.S. Patent he filed in 1983 for a "Method and Apparatus for Splitting Water Molecules[1]." In that, his 'water fuel cell' splits water into hydrogen and oxygen using an amplified audio frequency. The molecule is being constantly stretched and relaxed in time with the frequency applied, breaking the bonds and releasing the gas. What is important to note here is that very small currents are required to do this and Puharić publishes all his figures to back it up. The fuel cell consists of a metal cylinder, not unlike a standard spark plug, along with a standard screw thread and also features a spark gap.

Some of his many lectures can be found as YouTube videos, but it is now hard to find any details. It's difficult to find much information about him at all, even though the videos of lectures on YouTube clearly show him explaining in fine detail the inner workings of his inventions. That may be due in part to his acquaintance with the billionaire David Rockefeller. He was also an acquaintance of the Mexican president of the time. Both those acquaintances

Puharić converted an RV (Recreational Vehicle) to run on hydrogen sourced from water and drove it for thousands of miles across America.

1. U.S. Patent 4,394,230 1983 for a "Method and Apparatus for Splitting Water Molecules: https://patents.google.com/patent/US4394230A

could have propelled his inventions to great heights quite easily, but for one thing, the Rockefeller family made its billions in oil!!! Also, Mexico was the seventh-largest oil producer in the world as of 2006, producing 3.71 million barrels of crude oil per day, and was unlikely to support something that could effectively bankrupt them. Instead, his property was mysteriously burnt down, destroying much of his work. He also lived in fear as he claimed that the CIA made at least four attempts on his life. He had converted an RV (Recreational Vehicle) to run on hydrogen sourced from water and drove it for thousands of miles across America. Dr Andrew Michrowski[1], who had known him since 1975, told me that Puharić had travelled about 300,000 miles (about 500,000 km) in his RV (according to the odometer) whilst doing many trips from Ottawa in Canada to Mexico City using only available water (sometimes even packed snow and salt marsh water) as a fuel. He said that since he could travel so far on a litre or two of water, the gas tank would tend to show "empty" or very close to that.

He used one bedroom in the 3-bedroom motorhome as his "lab". The water fuel system not only powered the vehicle's 7-litre engine but also charged the motorhome's battery, enabling him to run the aircon and fridge. What is helpful in such a technology is that you only use the accelerator pedal to produce as much power-on-demand (the exact amount of oxyhydrogen) as you require when driving.

The last known location of this RV was on an estate in Devotion, North Carolina, a research haven for advanced scientific thinking, owned by Richard J. Reynolds III, grandson of R. J. Reynolds, the tobacco magnate. Andrija Puharić had been living and conducting research on the estate since 1980, joined by Dr Elizabeth Rauscher[2] (1937-2019) who was an American physicist and parapsychologist, with her husband—William van Bise, an engineer, lived for close to a year right next to it. Reynolds had allowed them to live there to conduct research into the effects of electromagnetic fields on brain waves. After Reynolds' death in 1994, the scientist said he had invited them to remain there as long as they wanted.

1. President of the learned society Planetary Association for Clean Energy (PACE) http://www.pacenetwork.org/ Which was co-founded by Andrija Puharić along with scientist Senator Chesley W. Carter and Dr Marcel Vogel (chief scientist at IBM) (Has archives with more information).
2. Prof. Dr Elizabeth Rauscher also authored an article about water-splitting in cars after due peer-review analysis as a physicist of Puharić's invention.

Where did his inspiration come from?

Puharić studied medicine as part of the Specialised Army programme and graduated with an MB in 1947. He completed his residency in internal medicine at Permanente Hospital, California. He was motivated to develop his system based on the analysis he did on a non-published item by Nikola Tesla on electrical resonance. Like Tesla, being of Yugoslavian descent himself and as an MD (with Physics and Electronic Engineering training) he had been asked by the Joint Chiefs of Staff in the Pentagon to study the Nikola Tesla papers before they were sent to Tesla's next-of-kin in Yugoslavia, to make sure that there were no uncovered technological benefits for the US military they'd missed.

This manuscript inspired him to use Tesla's insights for electronic amplification of sound (for hearing aids and for bugging systems) and upon reflection, further on, for the decomposition of the water molecule.

The electronics

Puharić, as an audiologist, used an *audio tone generator* (an oscillating electronic frequency) amplifying it to the required level. In this way, he was able to 'tune' the signal to the required 'resonant frequency' for optimum hydrogen production. In a vehicle, it would have to be constantly adjusted as demand changed along with road speeds/engine speeds.

Water at the microscopic level.

Puharić was observing the behaviour of water very closely, at a microscopic level. In one of his lectures, he talks about the high cost of using the equipment at Northwestern University to do this research. He spent many hours and a great deal of money analysing his experiments through a microscope and highly accurate test equipment. One thing he writes about in his patent (US 4,394,230 'Method and Apparatus for Splitting Water Molecules') are the orbitals of the water molecules. They are just that, 'orbits' that the electrons trace, along with the precise angles that combined atoms of hydrogen and oxygen hold under certain conditions. An atom is constantly spinning and this affects its appearance. Take, for example, a person spinning an object on a length of rope in circles around himself. As he spins faster, the rope would describe a flat disc through the air, which has a solid appearance. Fire jugglers use this principle to describe intricate patterns in the dark. Those patterns are orbitals. In the case of molecules, the orbitals are odd three-dimensional shapes due to the different charges of the component atoms and nuclei attracting and repelling each other.

The mechanics of a water molecule

Puharić shows diagrams of the structure of the water molecule and the angles between the two hydrogen and oxygen atoms. He then talks about how this setup is warped using the resonance until it breaks. The signal, in the form of a precise, gated 'step charge', has the effect of flexing it back and forth hundreds of times a second to weaken the bond. Why all this matters is that we may use a sound frequency in waveform to manipulate the shape of the water molecules so that the angles change in a predictable manner, flexing the angle between the two separate gases until the bond between them fails. It's almost as simple as that, except there are some other problems, for example, a build-up of gas on the water cell plates interrupting the process. Puharić, observing these issues, found that simply tapping the cell dislodged them. His patent describes all these things in detail.

Beware of the many variables.

It's an exact science requiring perfect geometry, plus, there are many variables to consider, such as temperature, pressure, and even precise timing. All of these need to be sensed and controlled if good results are to be achieved, which is easy to achieve with modern miniaturised electronics. If you are attaching this technology to a vehicle then all these variables are constantly changing. Heat, atmospheric pressure and the varying speed of the vehicle demand a constantly adjusting quantity of gas. He said in a lecture at the First International Symposium on non-conventional Energy Technology (PACE)[1] that **"This thermodynamic device has already been tested at ambient pressures and temperatures from sea level to an altitude of 10,000 feet above sea level without any loss of its peak efficiency"**.

When attempting to replicate this technology, care must be taken to consider

1. From the First International Symposium on non-conventional Energy Technology (PACE) "Cutting the Gordian knot of the great energy bind" (See Appendix 1)

all the finer details. For example, on the construction side, Puharić states that the *vitreous ceramic coating* is "highly technical" and "where the magic happens", so we can determine that it's not just an insulator! (I can only guess that the porous structure allows the fine hydrogen molecules through but nothing else). He also states that the 'critical gap distance' at the top of the 'thermodynamic device' is the 'known quenching distance for hydrogen'. Also, the 'nuclear spin relaxation time' of water is three seconds. Whatever does that mean? Well if you stretch or flex something, that's the time it takes for it to return to a state of equilibrium. That determines the exact 'gate' of the pulsed frequency. The precise frequency is of course also imperative along with its accompanying harmonics. Puharić says in his patent that the four harmonics produced by his circuitry each affect (agitate) the four apices of the tetrahedral form of the water molecule. The complex frequency produced is seen on an oscilloscope screen as a rippled triangular wave. (A combination of the carrier wave and its four harmonics).

Puharić's patent is very interesting in that he looks in great detail at what is going on at a molecular level. He even talks about the various shell configurations and orbital states of the atoms as energy is applied. That means exactly what happens to the water at various stages of applying an electric current. He lists eight phases, A-G. The first is just a 'dry test' to check the correct functioning of the equipment. Then as power is applied, he explains exactly what happens and why.

The thermodynamic device outlined is obviously a direct replacement for an internal combustion engine spark plug. Plus he mentions the use of an 'rpm' (revolutions per minute) sensor and an 'ignition control system' as part of the setup. I don't feel the need to quote large sections of his patent here, as it is currently freely available online (see Google Patents), (although at the rate I witness information on this subject disappearing, that may not always be the case!) But I will outline a few important features. He comments about the changing 'bond angles', that is, the molecular bond between hydrogen and oxygen. 104.45° and 109.5° are the two basic bond angles of water in classical-quantum physical chemistry. In the 'Summary of the Present Invention', he states "The present invention involves a method by which a water molecule can be energised by electrical means so as to shift the bond angle from the 104.45° configuration to the 109.5° tetrahedral configuration." (See my diagram on the right). Why? Because, quite simply, the constant flexing back and forth of the molecules at great speed (about five hundred times per second), will break the bonds, thereby separating the two gases hydrogen and oxygen. In his words *"Shatters the water molecule by resonance into its component molecules hydrogen and oxygen. The hydrogen, in a gas form, may then be used as fuel; and oxygen, in gas form is used as an oxidant".* ('Oxidant' means oxygen to help things burn. Hence coal fire flues, bellows, internal combustion engine air intakes! Conversely, if you cut off a fire's oxygen you can stifle the flame.)

Tetrahedral Structure of H_2O Molecule

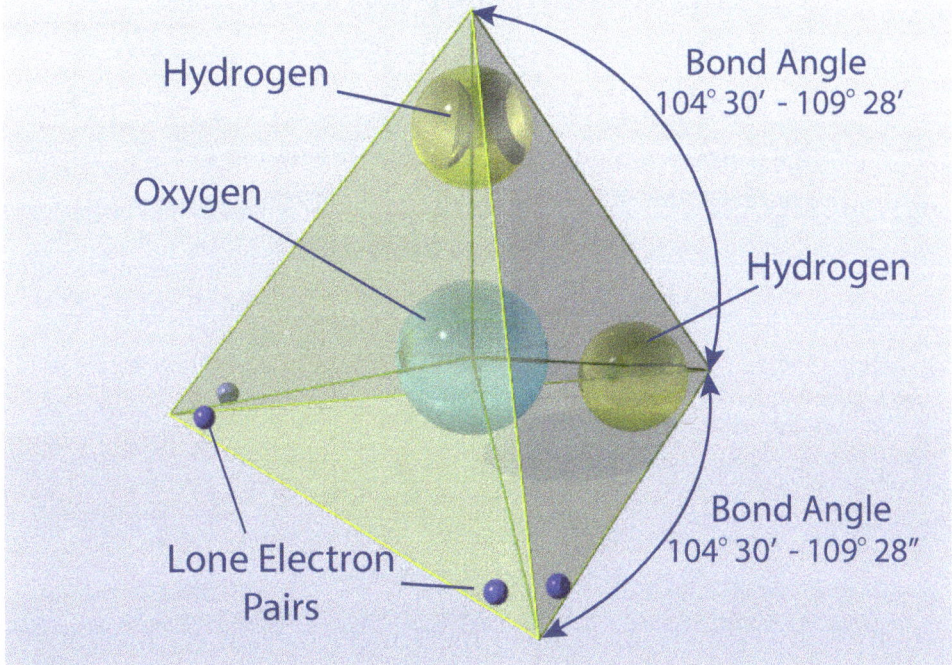

Illustration by the author, rendered in POV-Ray.
(Open-source Ray Tracing Application).

In his patent 'Method and apparatus for splitting water molecules' (US4394230A)', Puharić talks about the tetrahedral form of the water molecule and its 'orbitals'. The Oxygen atom has eight electrons, two pairs of which are shared with the two hydrogen molecules, the other two pairs are free. Each of these elements has an electromagnetic charge which tends to repel the other elements. That results in a tetrahedral form, albeit a little skewed. The ideal geometry of a tetrahedron has angles of 104.45°, but the molecular water version has angles of 109.28°. However, when energised by the magnetic field during resonant frequency electrolysis, within the pulsating electric and magnetic fields, the force pushing the electrons apart becomes stronger than that pushing the atoms away from each other and so the bond angle is forced back to 104.45°. It is the constant flexing between these two extremes that weakens the bonds and eventually breaks them.

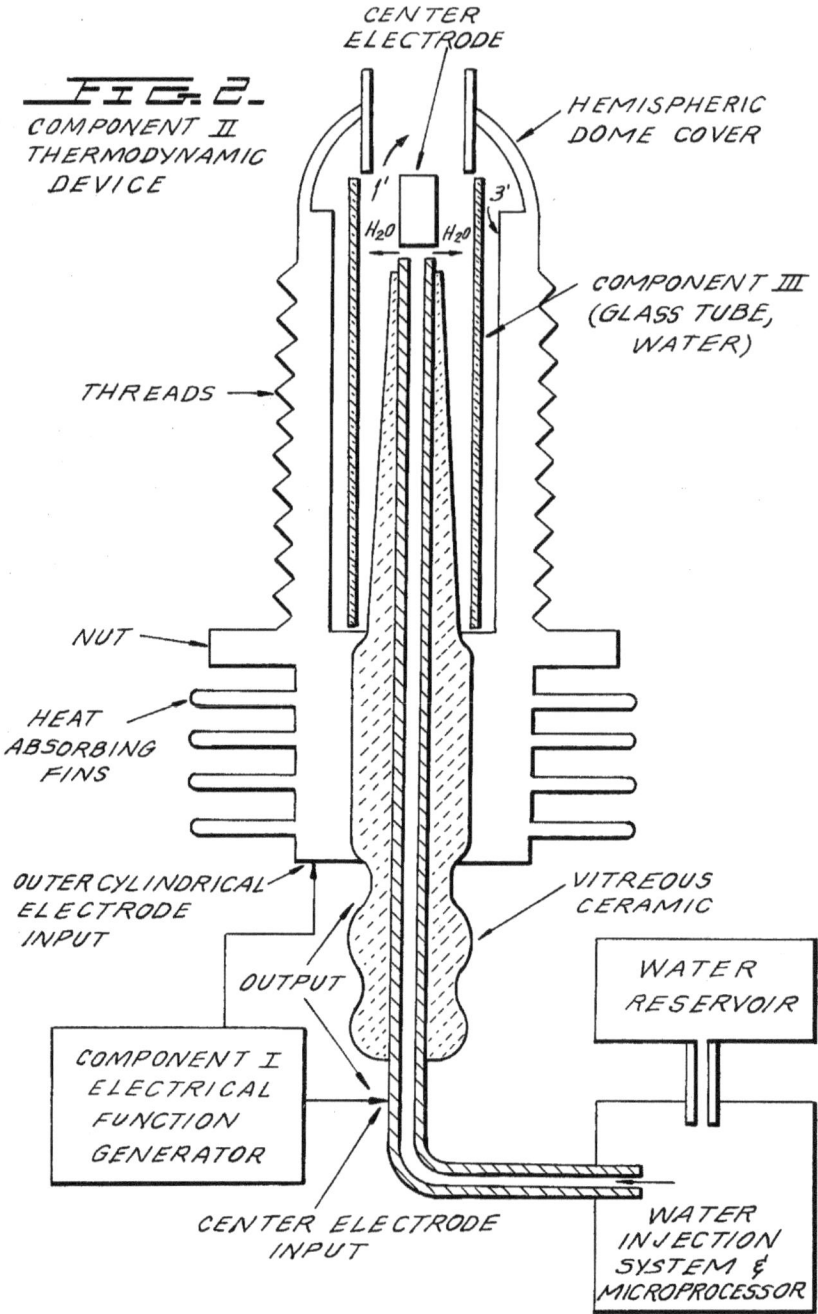

FIG. 2.

COMPONENT II
THERMODYNAMIC
DEVICE

CENTER
ELECTRODE

HEMISPHERIC
DOME COVER

COMPONENT III
(GLASS TUBE,
WATER)

THREADS

NUT

HEAT
ABSORBING
FINS

OUTER CYLINDRICAL
ELECTRODE
INPUT

OUTPUT

COMPONENT I
ELECTRICAL
FUNCTION
GENERATOR

VITREOUS
CERAMIC

WATER
RESERVOIR

CENTER ELECTRODE
INPUT

WATER
INJECTION
SYSTEM &
MICROPROCESSOR

A schematic illustration of the apparatus of Puharić's invention, from his patent including a cross-sectional representation of the 'thermodynamic device' that he calls 'Component II'

What is most interesting to me is that it only requires 1mA at 22v. Yes, that's just one milliamp! *Puharić is saying, in effect, that it takes just 1 mA to separate the water into hydrogen and oxygen!* Hardly any current at all. (Often, if you do something with finesse you don't need Brute force!). His final figures in the patent report that a 115% output compared to input is to be gained[1]. This is just an increased output due to the efficiency of reduced energy input. (If you don't understand that please study the figures on pages 11 - 15 of his patent).

This is not 'over-unity, 'free energy' or 'perpetual motion' as the energy is released from water, which is consumed in the process, just as any fuel is!

Puharić was a medical audio specialist. The invention seems to stem from a previous patent[2] (US3170993A), used to stimulate the body's facial nerves and blood. Indeed the electronic circuits detailed in that patent are the same. He uses an audio oscillator and audio amplifier to create the signal required for the resonant frequency (*"Means for aiding hearing by imparting modulated electrical signals to variable nerves of the facial system of a subject, comprising a radio frequency receiver for receiving transmitted signals corresponding to audible sounds..."*).

The science of water-splitting

For further information, included in Appendix 1, is a small, but important extract from Puharić's patent. It's naturally a little complex but does contain important general details.

1. He was thrilled when Golob Labs in New Jersey confirmed, as he suspected, that his system produced more energy than it consumed - which is at an average of about 112%.
2. Henry K Puharić 'Means for aiding hearing by electrical stimulation of the facial nerve system' Patent US3170993A: https://bit.ly/3x3gEV6

The highly respected scientist

On a final note, many sceptics discredit Puharić due to his interest in telepathy, parapsychology and the paranormal and even his research into water as a fuel contributed to a reputation as an unconventional scientist. However, it's surely quite unscientific to discredit someone because of their alternative interests. It's like discrediting Einstein because he refused to wear socks (yes, that's true!). Personally, I think that it's because anyone with an enquiring mind, particularly scientists, would be open to many diverse disciplines. I know that I am, which has resulted in the research for this book along with the understanding of all its extraordinary and unconventional principles.

Andrija was most noted in his lifetime for inventing a microelectronic hearing aid, but he also contributed to the fields of medical electronics, neurophysiology and biocybernetics. He made several important discoveries about the creation of life forms during electrochemical activity, including Chlorophyll life form, which led to the very important discovery of magnetite in blood as a carrier & coupler of external electromagnetic fields in the body.

During the late 1960's and early 1970's, his main occupation was his research work for Intelectron Corporation, a private organisation that he co-founded with investment funds provided by a group of New York businessmen. He maintained his role as President and Director of Medical Research at the organisation until 1971. During his time there, the organisation was awarded a contract with the United States Air Force Systems Command to further explore the electrostimulation techniques of hearing.

Andrija was extremely nervous that the US Navy might block any patent for his water-fuel invention for their stealth submarines. This is why he was sure that the Planetary Association for Clean Energy (PACE) as a Canadian Learned Society (at the Royal Society of Canada chambers in Ottawa) would be the first very public airing. It was also videotaped by a local cable vision studio, thus pre-empting potential extra-national closure - followed by their publishing of the proceeding as quickly as possible.

When Andrija died, the Planetary Association for Clean Energy (PACE) received a request from Rolls Royce aircraft engines in Atlanta as they were considering using his system for their future jet engines.

Why is the work of this man not common knowledge? As a doctor of medicine he made many important medical discoveries, especially relating to hearing and sound, in particular that if he focussed audio waves at a blood clot in an artery at blood's correct resonant frequency, it would break up and vanish. He then wondered what would happen if he did the same with water. He found the right frequency and the water easily broke up into its constituent parts of Hydrogen and Oxygen. He used sound waves. Fantastic. Now we are getting somewhere. Audio frequencies are sine waves, so the rate that the water molecules gave up their hydrogen and oxygen shot up from Faraday's out-of-date straight-line Direct Current results, which are too often quoted.

Phew! That was a bit heavy going, but fascinating, don't you think? Puharić used sound to split the water molecules. Using an audio amplifier is just one way of introducing a fluctuating resonant frequency. The Meyer twins which we will look at next used a different method based on a modified vehicle alternator.

Although that seems very different, in fact, the plates of the water fuel cell become like audio speakers. The resonant ringing can in fact be heard. You could realistically say that we are using sound to fracture the water molecule whatever the equipment used.

Some people have the nerve to call this pseudoscience. So, on to the work of Stanley and Stephen Meyer, the American twins who made great steps forward until Stan's untimely and mysterious death.

Stanley and Stephen Meyer were twins. Stan a mechanical genius and Stephen an electrical genius. We think Stan attended one of Puharić's lectures, but the idea went "Wow!" inside his brain. He was an inventor. Could *he* make a car run on Hydrogen from water? He homed in on alternating electrical current, as all cars already create that for producing sparks.

He experimented using alternating electricity (big volts, tiny current), two wave or three wave to increase the rate of fracturing the water even further. He got massive more Hydrogen. Plus loads of Oxygen to fatten the combustion mixture, combining that with exhaust gases perfectly matching the burn rate of existing fuel.

He didn't get Hydrogen from a petrol station tank and then try to store it like we do with petrol (hydrogen doesn't like it). He found it far easier to extract the hydrogen from the water right before it's used.

It worked. He converted a VW air-cooled car engine. Controlling the hydrogen quantity and managing the production enabled him to control a car's acceleration and deceleration safely just like we do with volatile gasoline.

He had found a world beater.

Then he was poisoned, and he died before he could release the concept to the rest of the world. Fortunately he patented his discoveries.

If Stanley could run his car on hydrogen that easily, so can we.

11

Stanley Meyer

born. August 24, 1940
died. March 20, 1998

- Was he murdered?
- Was he a fraud?
- A pseudoscientist perhaps?
- What happened to his water-fuelled car?

Stanley Meyer, genius engineer, original thinker and self-financed maverick inventor, with hundreds of patents to his name. He made one big mistake that cost him his life. He built a water-fuelled car that did not rely on costly oil-based fuel to function.

Stanley Meyer, from Columbus, Ohio, USA, submitted many hundreds of patents in his life, about 100 of which relate to water-fuel technology, all of which have now lapsed. He worked tirelessly on his water-fuelled, hydrogen-powered car for more than 15 years. It was a work-in-progress for which he sought more funding. Having refused a billion dollars from a middle-eastern oil cartel to stop work on it, at a meeting with potential financiers at a Cracker Barrel restaurant in Grove City, Ohio, according to his twin brother, he suddenly ran out clutching his throat and exclaimed "They poisoned me". The official verdict was that he died of a brain aneurysm, but would that cause you to run around shouting? I think not! but more on that later.

Stan Meyer's Water Fuel Cell

Meyer's cell produced far more hydrogen/oxygen mixture than could have been expected by simple electrolysis. According to Meyer, the device required less energy to perform electrolysis than the minimum energy requirement predicted by conventional science.

A demonstration made before Professor Michael Laughton, Dean of Engineering at Mary College, London, Admiral Sir Anthony Griffin, and Dr Keith Hindley, a UK research chemist apparently confirmed this claim.

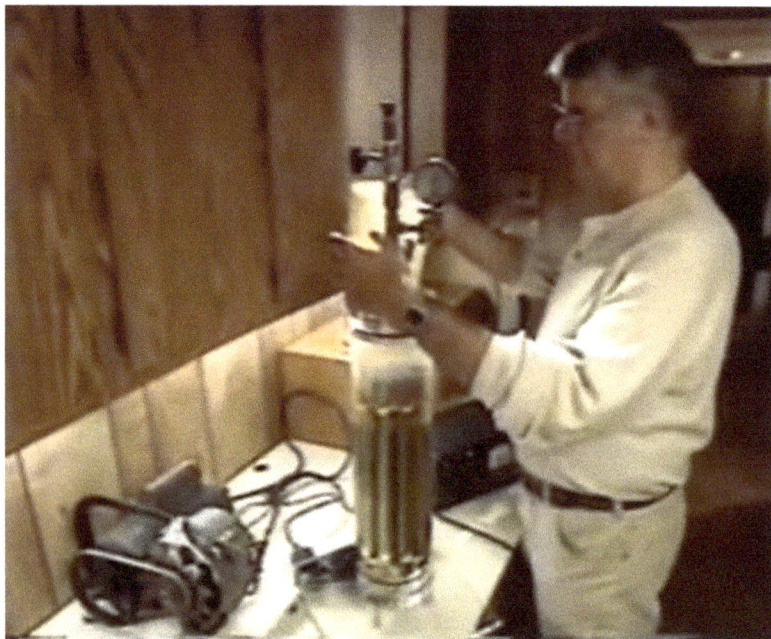

Stan Meyer in his lab with the fuel cell that he demonstrated during his successful patent application, with the alternator providing the AC power. It only required 1/2 Amp current at 12 Volts to produce significant quantities of hydrogen.

Stanley Meyer's WFC (Water Fuel Cell)

After the sad demise of Stanley Meyer, his personal effects, including his water-powered car and associated experimental equipment, became available for close scrutiny. Seen below, next to his patent drawing, is one of many photographs of his Water Fuel Cell that was accepted for US patent[1]. Photographic archives exist of many other items. A team of people are actively deconstructing every item meticulously to discover their secrets and answer outstanding questions.

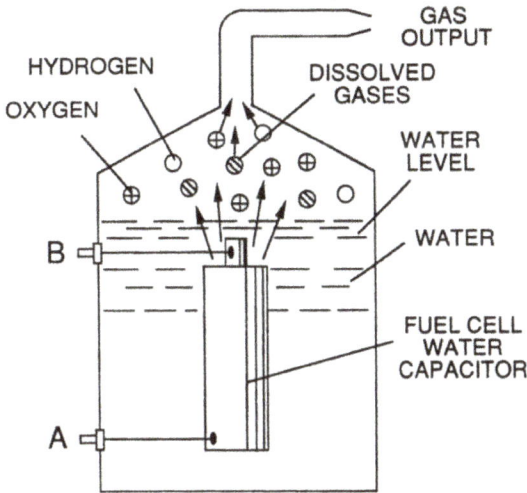

Stanley Meyer's 'Water Fuel Cell' (wet cell), an electrolytic cell consisting of a set of double stainless steel tubes in a water container. The inner tube of each acting as the Cathodes and their outer tubes the Anodes.

1. Original Meyer Water Fuel Cell c.1990:
   https://www.youtube.com/watch?v=nOGAkRkCWfA

Stan Meyer's water-fuelled car

Along with his surviving twin brother, Stephen, Stan Meyer designed and built the 'Water Powered' car, based on a Volkswagen beach buggy, incorporating the resonant frequency principle that I introduced earlier. The car is a VW beach buggy with a 'flat four' (horizontally opposed, four-cylinder air-cooled 1600cc internal combustion engine). As a simple open-top configuration, it was an ideal platform for experimentation, with plenty of room to bolt on the banks of control circuits and extra parts required. The VW flat-four engine is renowned for its simplicity and allows easy access and plenty of space for modifications. Since his death, the car and many other items from his estate have been located, and the new owner reverse-engineered it to try to uncover its secrets.

Firstly, despite his claims, I don't believe it was 'his' idea. He started work on it shortly after Andrija Puharić lodged his very similar patent[1], but with certain key differences. He did have his own novel approach though, with a few genius innovations added. But by his own admission, it took longer than he thought, about 15 years in fact, and still there were bugs to iron out. Although it attracts a lot of interest, the car was not, in fact, his crowning glory. That honour goes to his scaled-down 'retrofit kit' which will be covered in the next chapter, and for that reason, I'm only going to cover the basics of how the car worked.

In his own words. *"This is the pre-engineering unit in order to satisfy the U.S. code of operability on section 35 section 101 and it was developed as a pre-engineering unit to show the operability of all the different operational parameters of the water fuel cell"* and *"The Hydrogen Computer System will be **miniaturized down to several IC chips**"* which *"will allow us to give the economics to apply to a conventional car."* it was only intended to meet the demands of his U.S. patent applications.

There are videos currently available on YouTube where he explains how it all works, demonstrating various components working on a test bench and on the car. One such video ('Stanley Meyer's water car's first run'[2]), shows where Stan and his twin brother Stephen, get the car running solely on water, with the original fuel tank completely removed from the vehicle.

1. Patent US 4936961A ('Method for the Production of a Fuel Gas' or 'water fuel cell' June 26, 1990): https://patents.google.com/patent/US4936961
2. Stanley Meyer's water car's first run: https://www.youtube.com/watch?v=Z5afwEcZ3Ok

"This is the pre-engineering unit in order to satisfy the U.S. code of operability on section 35 section 101, and it was developed as pre-engineering to show the operability of all the different operational parameters of the water fuel cell"

~Stanley Meyer.

The Meyer brother's water-fuelled, hydrogen-powered car, with a Volkswagon air-cooled, flat-four, 1600cc engine. The ideal platform for experimentation. Resonant-frequency electrolysis is at the heart of its operation.

His car and many other items from his estate have been acquired and reverse-engineered by several people who have been experimenting and reproducing the parts (which can now easily be obtained as STL files for 3D printing or CAD, see open-source-energy.org). There are also all his patents, although there is some thought that he was holding a few things back in order to gain market advantage. His car was a rear-engined VW beach-buggy style car, making work on it easy, but the electronics of the day were cumbersome, although he himself, in one of his videos, says that when it is all finalised the electronic control systems could be shrunk down to a fraction of the size. How right he was, as we now have a glut of SBCs (Single-board computers) the size of a mere pack of playing cards. A 'Control Unit' which he called the 'GMS' ('Gas Management System'), consisting of a server rack containing 18 module circuit boards, mounted on the dashboard, controls the system, and various other components added to the car are clearly visible. In his words

"..the Hydrogen Computer System which was designed in order to be able to process the fuel to produce the hydrogen gas from water and do it economically and be able to control its firing allowing the Volkswagen engine to run off hydrogen."

The main principle of the car is that water in the fuel tank is fed into a 'water fuel cell' where it is subjected to a resonant frequency which breaks down the water molecules into hydrogen and oxygen gas. This gas is then burnt in the combustion chamber instead of fossil fuel vapour.

The Water Fuel Cell from Stan Meyer's car. Within it are eleven double stainless steel tubes.

The Control Units which were fitted to Stan's prototype water-fuelled car. He said that "when it is all finalised the electronic control systems could be shrunk down to a fraction of the size".

A key component of his design is a special kind of coil. A 'bifilar' coil, designed and patented by the great Nikola Tesla, has a few interesting special properties. This boosts the voltage in a particular way and along with the circuitry, Stan was able to break the hydrogen-oxygen bonds of the water flowing through the system. In his lectures, he states that it is the collapsing magnetic field, with *very little amperage* that does this. His circuit produces what is called a 'step-charge' voltage, which is a gated, pulsed frequency that builds up and then splits the molecular bonds apart. This is very precisely controlled and needs to match the resonant frequency of the water. He incorporated a device to control the acceleration, or flow of water, matched with the correct frequency.

1. If you need more detailed information, see Alex Petty's wonderful 'Energy Research Journal' site www.alexpetty.com 'Mappings Between Meyer's Hydrogen GMS Cards and Patent Schematics', as well as 'Stan Meyer explains the (Water Fuel Technology) on the dune buggy (https://www.youtube.com/watch?v=EJ-zb0CmGag&t=209s) from May 1992', an outline and transcription of Stan Meyer's video. (note: I have also downloaded all the material referenced here in case it gets taken down).

Let's take each one in turn here to briefly explain its function. The patent 'A Control and driver circuits for a hydrogen gas fuel-producing cell'[1] details the operation and schematics involved. (Also refers to the previous patent 'Method for the Production of a Fuel Gas' or 'water fuel cell', remarkably similar to the aforementioned Andrija Puharić's patent). An innovative 'Laser distributor system' which is designed to fit between a conventional distributor cap and its housing, sends signals to the GMS.

One very clever innovation was the answer to the problem of the hydrogen gas igniting instantly, which is much too soon for the engine's normal firing cycle. This was the electronic regulator that mixes non-combustible exhaust gases, to retard the combustion of the hydrogen. This, mixed with ambient air, through an 'Electronic Injector system regulator unit' meters and mixes combustible and non-combustible gases to provide an accurate burn rate control in the correct proportions to allow fine control of the combustion process. There is then the 'Gas Processor Unit', which ionizes the ambient air to kickstart the water fracturing process. On top of the water tank, we have the 'Resonant Cavity Unit' which subjects the water as it is drawn up to a very

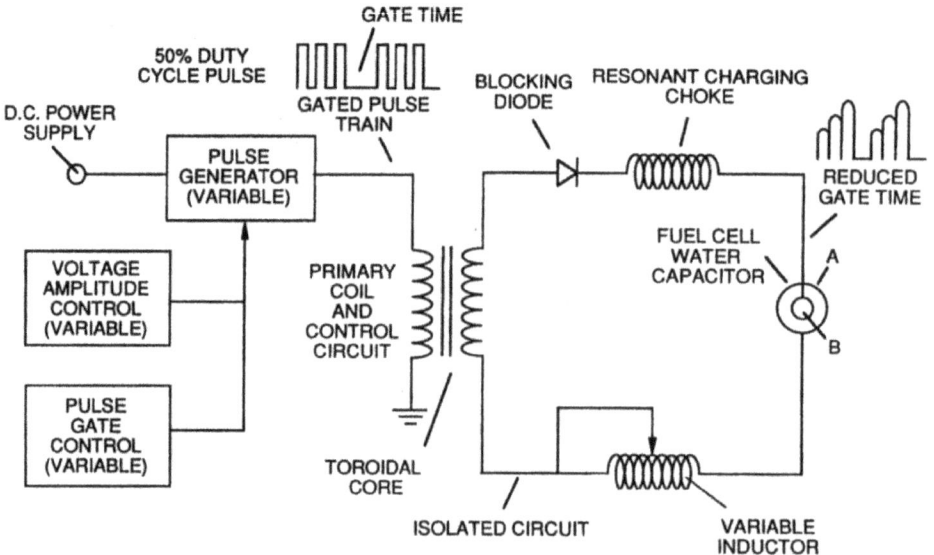

One of Stan Meyer's circuits shows how the resonant frequency is applied to the Water Fuel Cell. (There are a few iterations, such as the 8XA and the 9XA/9XB), as he continued his research, all based on an L/C inductor/ capacitor oscillator principle, as seen in chapter 7.

1. Patent US WO 92/07861 (14th May 1992 'A Control and driver circuits for a hydrogen gas fuel-producing cell'): https://bit.ly/3GDPTL7

high intense pulse voltage field and restricts the amps, initiating the electrical polarisation process. The fluctuating voltage of the resonant frequency, matched with that of the water, is allowing the release of the hydrogen economically from water. This is where the rate of the production of the hydrogen gas on-demand is controlled in response to the engine's requirements, which will vary considerably on a continual basis as the vehicle accelerates and decelerates, and changes gears.

Stan Meyer's VIC (Voltage Intensifier Circuit)

This was key to Stan Meyer's water-fuelled car and other patents. Why? Because in the process of efficient hydrogen production we need a large voltage but a relatively low current. Stan's VIC, or 'Voltage Intensifier Circuit' boosted the voltage with the use of coils of many turns of copper wire. The important thing to note is that it's the voltage that does the work, not the current. The amps are in fact actively restricted. A bank of these 'VIC' coils is fitted near the engine so that one set is matched to each of the nine resonant cavities of the fuel cell, which are concentric stainless steel tubes arranged in a circle within the water fuel cell. (As detailed in patent US WO 92/07861 (14th May 1992 'A Control and driver circuits for a hydrogen gas fuel-producing cell').

Stan Meyer's VIC (Voltage Intensifier Circuit) Coils. "The key to deriving oxyhydrogen gas from water whilst consuming almost no power".

Another item that fits within the water tank is the 'Steam Resonator', the purpose of which is to prevent the water from freezing if the ambient temperature should drop so low. What is particularly interesting about this, is that it can be used to heat water with minimal current, by using a similar 'resonant frequency' technology, in other applications such as for hot water in the home (see also page 88, 'Ohmic Array Technology').

The basis of most of Meyer's work is the 'LC Tank circuit'. (See section on inductors (Chapter 7, p78). That is a basic oscillator circuit which provides the resonant frequency required to split the water into hydrogen and oxygen. Of course, here we have a set of coils and the water fuel cell itself is the capacitor, with its anode and cathode plates or tubes.

Resonance of a tube

The resonance of a tube is related to the length of the tube, its shape, and whether the ends are open or closed. There are two types of tubes, those that are open at both ends and those that are open at one end and closed at the other. Of course, many musical instruments are made of tubes that are conical or cylindrical and like strings, vibrating air columns in ideal cylindrical or conical pipes resonate at a particular frequency due to their length and internal (physical) volume. Stopping finger holes with your fingers or mechanical pads has the effect of lengthening or shortening the effective column of air. Acoustic resonance results in amplified sound waves whose frequency matches one of its own natural frequencies of vibration (its resonant frequency).

Tuning resonance with a tube slot

You may have noticed in photographs and videos of Stan's tube sets that the outer tubes have a rectangular slot cut in the top. This is for the same reason that church organ pipes usually have slots at the top of the pipe, and that is to raise its pitch or frequency of vibration. It's a way of fine-tuning the outer pipes in the Meyer cell which will otherwise resonate at a lower frequency than the inner pipes. Therefore, the slots cut by Stan are to raise the resonant frequency of the larger pipes, to match the resonant frequency of the inner pipes. If he didn't do this, the pipes would have to be different lengths in order to have matched frequencies. One way of tuning a pipe is to hang it up on a piece of thread and tap it so that it will produce the resonant pitch of the pipe. Cutting a slot in the end, suspending it on a piece of thread and tapping it, will allow the pitch of the two pipes to be compared. The slot can then be gradually enlarged until the two tubes match. Another important point to watch is that mounting the tubes too firmly can have the effect of dampening the resonance. Notice the spring-loaded base mountings on Stan's Water Fuel Cell (page 139)

Did Meyer copy Puharić?

Because the two men (Meyer and Puharić) did not work together, Meyer said that it took him 15 years to perfect the idea. Longer than he thought it would. Particularly, I think as he went about it his own way. A bit like re-inventing the wheel.

Meyer's approach was slightly different from that of his possible mentor Puharić, an audiologist who used a tone generator and amplifier to produce the signal that controlled the breakdown of the water molecule. Meyer, however, used existing automotive systems including the vehicle's alternator, which already outputs an alternating current and his own special high voltage circuit to create a spark that ignites the fuel gas in the cylinders.

One thing which makes me believe that Meyer copied Puharić's ideas is that when he developed his retrofit kit to suit any vehicle, he used a direct fuel injector instead of the tube arrays that he'd used previously for an Electrolyser. The one Puharić described in his patent looks like a direct replacement for a spark plug, with its threaded body and ceramic insulator, except that it creates hydrogen on-demand within it. In the same way, Meyer claimed that his design which "directly replaces the spark plugs" was "where the magic happens".

Nevertheless, Meyer and Puharić had one thing in common, they were both humanitarians who wanted the best for everyone. They could see the wider implications of a clean, environmentally friendly source of power. Perhaps the biggest test of his resolve came when a middle-eastern oil cartel offered him a $Billion to cease working on it, which he refused.

Stanley Meyer's mission to help mankind probably stemmed in part from his being a devout Christian (He had a sticker (decal) on the car saying "Jesus Christ is Lord") and he claimed that he received his inspiration 'from God', but I personally believe that wasn't true. In the course of my research, I found that for a short time he had attended lectures by Andrija Puharić at Northwestern University. They were also geographically in close proximity. It would be far too much coincidence for the two to be talking about the same extraordinary thing without Puharić being the initiator of the ideas. However, since Puharić was highly interested in what some would consider 'the dark arts', such as telepathy and psychic healing it wouldn't have gone down particularly well with Meyer's Christian ideals, to say the least.

There are many YouTube videos of his lectures and construction of the car as well as many patents available online (see Appendix 2). In case any individual or government department has any ideas about removing them, I have downloaded them all and am freely sharing them as part of my open-source project.

If anything untoward should happen to me, as it did to Stanley Meyer, this time, it wouldn't prevent this technology from getting out to the wider world. In fact, it would have the opposite effect. I would become a martyr, my mission complete and the world asking questions.

The sham court case

Many sceptics love to mention the sham court case on September 3, 1996, Common Pleas Court, in Fayette County, Ohio, claiming that it proves that Stan was a fraud. However, who in their right minds would expect the inventor of a water-fuelled car to get a fair trial right in the middle of a state where oil is the main industry? (The headquarters of the "Standard Oil Company of Ohio" is just 150 miles away in the same state! See map in Chapter 14).

He was being sued by two investors to whom he had sold dealerships offering the right to do business in Water Fuel Cell technology. (To me, that shows just how advanced his water-fuel technology actually was). He was not even allowed any witnesses to help testify in his favour. The technicians that were present obviously did not know what they were looking at and Stan had to warn them during the trial as they filled up his fuel cell with water and prepared to turn it on, that what they were doing was dangerous due to the amount of hydrogen building up. He said that the extremely high electrical potential generated by the coils was potentially lethal and so he wouldn't be prepared to accept any liability for the consequences. The court then did not allow the test to proceed, calling that a "lame excuse"! The court Judge, William Corzine III, subsequently found him guilty of 'gross and gregarious fraud' for selling dealerships to investors based on technology that did not function as claimed and for this serious offense was ordered to pay those investors back and pay a $1 fine (Yes, you read that correctly!).

Meyer intended to take his case to the Supreme Court of Ohio because he felt there had been judicial misconduct. He outlined his grievances in a "Public Notice to Inform" on December 20, 1996. He also wrote formal complaints to New Energy News and the London Sunday Times in regard to articles about the guilty verdict of the Ohio Court. His "water fuel cell" was later examined by 'three expert witnesses' who found that there "was nothing revolutionary about the cell at all and that it was simply using conventional electrolysis." That is of course complete nonsense as even a layman could plainly see by reading the patent that it functions with a resonant frequency which conventional electrolysis plainly does not. (If you have understood my book so far, you too would understand how preposterous their statement was).

In an interview with Max Miller ('Meyer according to Don'[1]), Don Gabel said that a court worker who was present at the proceedings told him that he'd "never seen anyone 'so stonewalled and railroaded'". Stan's surviving twin brother Stephen said in a radio interview with Jas Robey[2], "It was not about the car at all". It's certainly not proof that he was a fraud or that his water-fuelled car or any other of his ideas didn't work, as many people seem to think.

Was Stan Meyer a 'conman'?

If so, he was the worst conman in history. Surely by their very nature, conmen are lazy and try to benefit from other people's efforts. Stan Meyer was the very opposite of that. He produced thousands of patents and worked tirelessly for over 15 years to produce a working prototype of a water-fuelled car and other outstanding inventions and patents. No, you only have to look closely at what he created to see that he was most certainly not a conman.

His sad demise

Meyer needed to raise more funding to get his invention to market,

1. 'Meyer according to Don' (Gabel), (Race tuning expert and custodian of Meyer's car): https://www.youtube.com/watch?v=PinYM3AzGaA
2. Jas Robey BlogTalkRadio Stephen Meyer interview, https://www.blogtalkradio.com/waterfuelmuseum/2007/03/25/stephen-meyer-part-1

particularly his new retrofit kit, adapting any vehicle to water-fuel, which is why he approached the investors who were from Belgium. The day Meyer died he was about to have lunch with his twin brother Steven Meyer and the two Belgians at a Cracker Barrel restaurant in Grove City, Ohio. (Oddly on Stanley Meyer's death certificate the men are called "NATO officials" - which they were not).

Did Stan Meyer die of an aneurysm? (As in the official report).

Some things simply don't add up.

After they made a toast with cranberry juice, Meyer ran outside, vomited and, according to Meyer's brother Steven, said *"They poisoned me!"*, while clutching his throat, but, the toxicology reports did not find any evidence of poisoning.

The Franklin County, Ohio, coroner's report found that his immediate cause of death on March 20, 1998, was a "rupture of cerebral artery aneurysm." However, aneurysm symptoms include a sudden agonising headache which has been described as a "thunderclap headache", similar to a sudden hit on the head, resulting in a blinding pain unlike anything experienced before. Would that make you want to jump up and run around? I think not. Most likely, you would drop to the floor incapacitated. Also, the number of aneurysms that actually rupture is quite small, with about 3 in 5 people who have a sub-arachnoid haemorrhage dying within 2 weeks. (i.e. not immediately). Although vomiting is a symptom, I doubt that anyone with such a condition would be running outside "Clutching his throat" when the massive headache is the primary cause.

Grove City, Ohio, police investigated and did not find any evidence of foul play, but immediately after Stan died there was a murder investigation. If the death was a suspected aneurysm, why would that even have been a consideration? Has anyone ever murdered anyone by subjecting them to an internal brain condition? , I wonder. I very much doubt it.

In my opinion, he was poisoned to suppress the water-fuel technology and that oil companies and the United States government were jointly involved in his death as they both stood to lose a great deal if the technology replaced fossil fuels. After all, the average salary for coroners in the United States is around $70,000 per year. It's surely not beyond reason to see how a relatively small bribe could determine the contents of a coroner's report, in my opinion.

Where is the car now? What happened to the car after Stan's death?

A man called Dave Holbrook ended up with the car, being named in Stan Meyer's widow's will. So who exactly is he? And how did he end up with the car? His father, Charlie Holbrook *who was originally a sceptic*, took it upon himself to prove that Stan was a fraud and his car was a con. Setting out to disprove it for himself he visited Stan's home in person. The outcome was the opposite. He was won over by Stan, ended up working with him and became a great friend of the family. Indeed his wife and Marylin, Stan's wife, also became great friends. So much so, that after Stan's untimely death, the car was passed on to him in Marilyn's will, despite offers of large sums of money being offered for it by third parties. Charlie only had it for about two years before he himself died and the car ended up in the possession of Charlie's two sons, one of which was Dave Holbrook.

Charles was present on many occasions when the car was running, but despite that, nobody has been able to get the car back to its original working condition. There are a number of reasons for that. Firstly, it has been difficult to re-assemble the car which had been dismantled. Secondly, many parts were removed by Stan as he was in the process of refining his retrofit kit (see the next chapter). Thirdly, while Stan's possessions were in probate for four years, parts were removed and other goods including a safe full of papers and computer disks were taken away by the police and government agencies. Some parts were removed "For security", in case the car was stolen since it was attracting so much interest and had considerable monetary value.

The mystery of the alternator

So for an example of how parts came to be missing, Stan had significantly modified a stock Ford Alternator with a different winding and internal connections as well as removing redundant components that didn't work with the rest of the system. The modified one was adapted to take 110 volts, whereas the original, like most alternators, was only rated for 12 volts as any more than that would melt the wire within the coil. In Stan's version, for example, you don't need the regulators or the capacitors. However, the original had been burnt out and unwittingly replaced with a stock Ford item. That is because, Charles, the new owner, did not fully understand all the intricate complexities of it and while trying to start it after Stan's demise, he accidentally burnt out the highly modified alternator. He took it to the repair shop which simply replaced it with another stock Ford alternator. (100 amp, which has six diodes for a full-wave bridge rectifier for three-phase).

Issues like that don't help piece the puzzle together at all. However, people are actively working on the problem, such as Daniel Donatelli of Secure Supplies (a member of the open-source-energy.org forum). Here's a link to a 24-minute YouTube video[1], 'Alternator secrets' where he gives a detailed description of the mods that you need to do to a stock alternator, to replicate Stan's.

Meyer's patents

Three years after his death, many of Meyer's patents, as part of his estate, were registered in the name of his widow, Marilyn Meyer. The registrations on Meyer's patents eventually lapsed in 2007 and the patents are now freely available in the public domain. It's not currently difficult to find these patents, although that may not always be the case.

Despite all that, there are a few dedicated groups of people who are currently trying to replicate the car based on what is left in an attempt to get it working again. But for me, having researched it all for so many years, I realised that his retrofit kit was far more advanced, can be retrofitted to any engine and would be much more useful to resurrect. We'll be looking at that in the next chapter.

Also, Stan's surviving twin brother Stephen, who was the brains behind the electronic systems, has also gone way beyond what Stan achieved. I'll explain all about his work in chapter 13.

---

1. Alternator secrets (analysing Stan's alternator mods) 25-minute YouTube video. (interspersed with self-promotional clips) Secure Supplies are members of the open-source-energy.org forum and they also sell parts and completed Stan Meyer components for you to try experimenting yourself. https://youtu.be/7tGTCNGfIrk

Relevant patents granted

(Can be found online and also on HYDODproject.blockchain see page 271)

USP # 4,936,961 – Method for the Production of a Fuel Gas
USP # 4,826,581 – Controlled Production of Thermal Energy from Gases/
USP # 4,798,661 – Gas generator voltage control circuit
USP # 4,613,779 – Electrical Pulse Generator
USP # 4,613,304 – Gas Electrical H Generator
USP # 4,465,455 – Start-up/Shut-down for H Gas Burner
USP # 4,421,474 – H Gas Burner
USP # 4,389,981 – H Gas Injector System for IC Engine
USP # 4,275,950 – Light-Guide Lense
USP # 3,970,070 – Solar Heating System
USP # 4,265,224 – Multi-Stage Solar Storage System
USP # 3,970,070 – Solar heating system .

Additional points

- For US patents to be granted under Section 101, Meyer's experiments, would have to be demonstrated to work as specified. The successful demonstration of the invention to a Patent Review Board is a requirement.

- An Ohio TV station showed Meyer successfully demonstrating his water-fuelled, hydrogen-powered car. He estimated that it would only need 22 US gallons (83 litres) of water to drive from Los Angeles to New York (a distance of 2, 446 miles/3936 km).

- As ordinary tap water is non-conductive, it normally requires an electrolyte to be added, (e.g sulphuric acid, potassium hydroxide, etc.) to make it conductive. However, Meyer managed to achieve great efficiency with just pure water and no additives.

- Excess energy input actually produces heat, but Stan Meyer's fuel cell remained cold, even after hours of use.

Stan's documented results

Any good scientist will publish his results for comparison and Stan Meyer was no exception. On the next page, I have included some of his figures which are astonishing. They show how using *voltage* and *pulsed electrolysis* to disassociate water into hydrogen and oxygen yields much better results than Faraday, who was mainly using *current (amperage)*.

WFC Gas-Yield Vs Prior Art Electrolysis (*issued Sept. 14, 1976*)

Inoue U.S. Patent No. 4,184,931; Horvath U.S. Patent No. 3,980,053

Continuous Electrolysis:
55 cc/min. @ 1 amp (Gases comprise 15%; Moisture 85%)
20% Potassium Hydroxide Electrolytic Aqueous Solution (55 cc/min. @
1 amp) X 60 min. X 4.0 amps consumed = 13,200 cc/hr. 13,200 cc/hr. -
[11,220 (85% Moisture Content)] = 1,980 cc/hr.

Pulsed Electrolysis: 20 Microseconds 50% Duty Pulse Rate
68 cc/min. @ 1 amp (Gases comprise 97%; 3% Moisture Content)
20% Potassium Hydroxide Electrolytic Aqueous Solution
(68 cc/min. @ 1 amp) X 60 min. X 4.0 amp consumed = 16,320 cc/hr.
16,320 cc/hr. - [(489.6 (3% Moisture Content)] = 15,830 cc/hr.

Meyer Tubular-Array Fuel Cell: Photo Exhibit (E2/E2A): U.S. Patent No. 4,389,981

WFC Voltage Disassociation of the Water Molecule

194.2 cc/min @ 1 amp (Gases comprises > 99.99%; ≤ 1 ppm Moisture
Content) Tap Water: [Potassium (1ppm) / Sodium (10 ppm)] Contaminates
(194.2 cc/min. @ 1 amp) X 60 min. X 4.0 amp Leakage = 46, 600 cc/hr.

Therefore, WFC Differential Efficiency Factor:

194.2 cc/min. - 68cc/min. = 126.268 1.85 or **185% WFC improvement
over prior art**

Application Note:

Faraday's Laws of Electrolysis provide that in any electrolysis process the
mass of substance liberated at an anode or cathode is in accordance with the
formula

$$m = zq$$

Where m is the mass of substance liberated in gram's, z is the electrochemical
equivalent of the substance, and q is the quantity of electricity passed, in
coulombs. An important consequence of Faraday's Law is that the rate of
decomposition of an electrolyte is dependent on current and is "independent
of voltage". For example, in a conventional electrolysis process in which a
constant current i amps flows to t seconds, $q = it$ and the mass of material
deposited or dissolved will depend on i "regardless of voltage." provided that
the voltage exceeds the minimum necessary for the electrolysis to proceed.
For electrolysis the Decomposition Voltage Potential is very low (Typically
1.55 - 2.25 Volts). (*See Case Institute of Technology, Fundamental Principles of
Physical Chemistry, Electrolysis Decomposition Potential Chart.*

12

Stan Meyer's retrofit kit
(For virtually any engine!)

*"50 plus patents come together
into one singularity"*

*"The installation is a very small, lightweight,
compact electronic control system"*

Stan Meyer.

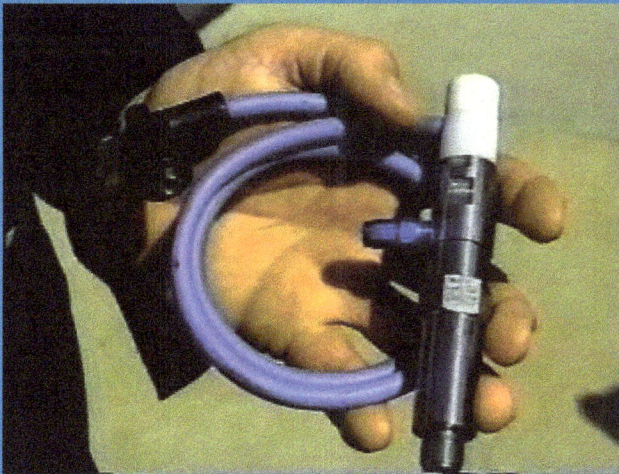

*Stan holding one of his WFC (Water Fuel-Cell) Injector
Plugs, the key component of his retrofit kit.*

Stan's 'retrofit kit' that had the potential to easily and inexpensively convert *any* internal combustion engine to water fuel could have far-reaching consequences for the planet but was prematurely cut short. That story is here.

S tan Meyer's car was a test bed for ideas and helped to set up the correct parameters for efficient resonant frequency electrolysis using all the circuitry and test equipment on board. However, as he said, it was ultimately just a pre-engineering unit built *"in order to satisfy the U.S. code of operability on section 35 section 101"*. His main goal was to refine everything down to one simple set of components that could be retrofitted to *any vehicle* with an existing internal combustion engine. He already had it in the final pre-production stages including a price of *"$1500 for a car and about $5000 for a truck"*. He said that *"50 plus patents come together into one singularity"*, meaning the fuel injectors that incorporated all the resonant-frequency water-splitting technology. It is also important to note that there are no exact plans available for the correct installation of all the parts, as far as anyone knows. Although there are patents and schematics they serve different purposes and leave big questions about the exact assembly and setup parameters. Unfortunately, one thing is certain, essential backup documentation is definitely missing.

The retrofit kit consisted of half a dozen fairly inexpensive peripheral components that could be easily manufactured. The Water Fuel Injectors simply replaced the existing spark plugs and a laser distributor went in between the existing distributor and its cap. The alternator needed modifications or swapping out as there were internal components that were not required. The existing fuel tank could be used if it was cleaned out and simply filled with water. No extra chemicals needed to be added.

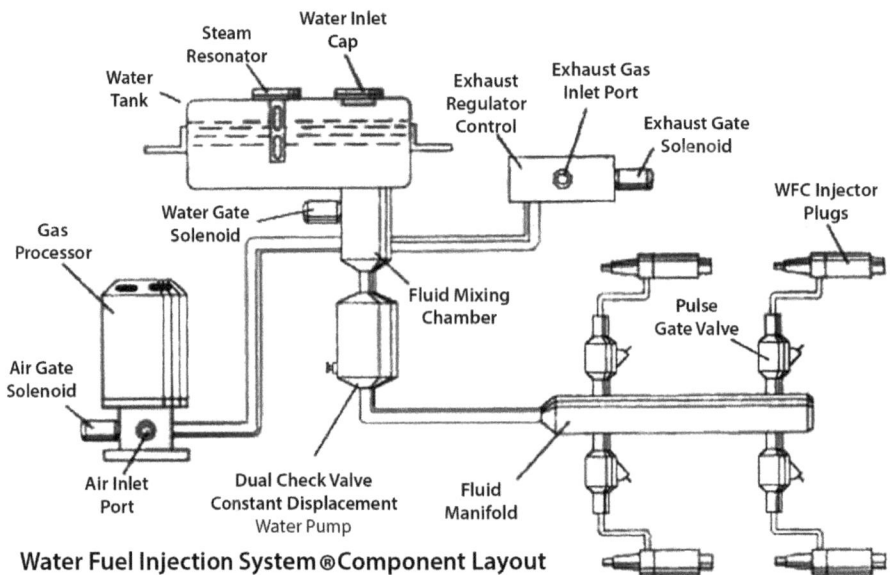

Water Fuel Injection System®Component Layout

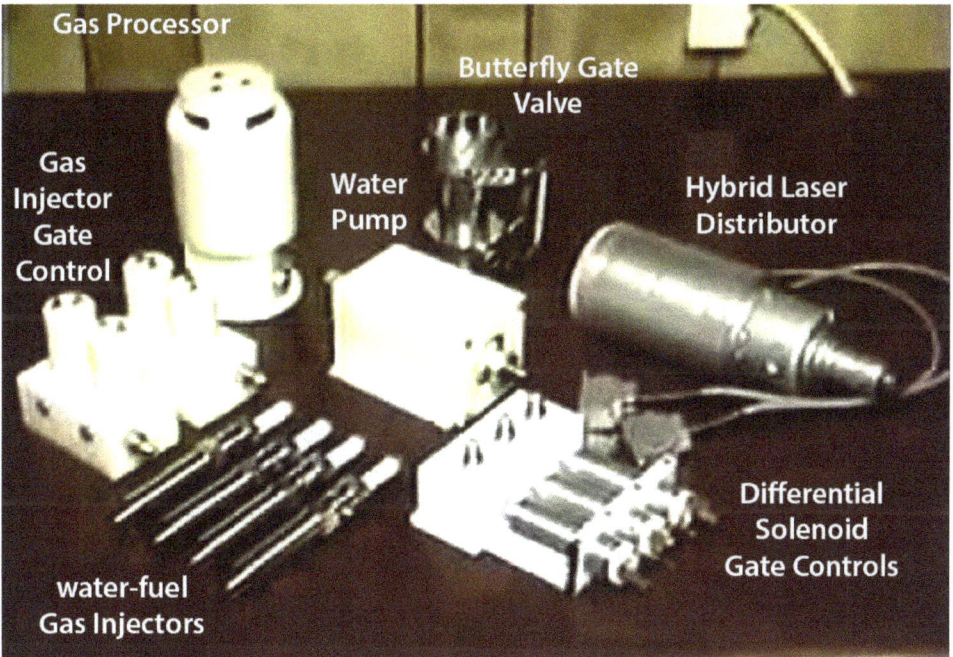

*Stan Meyer's retrofit kit photo of the actual parts (Above),
and where they fit on the engine (Below).*

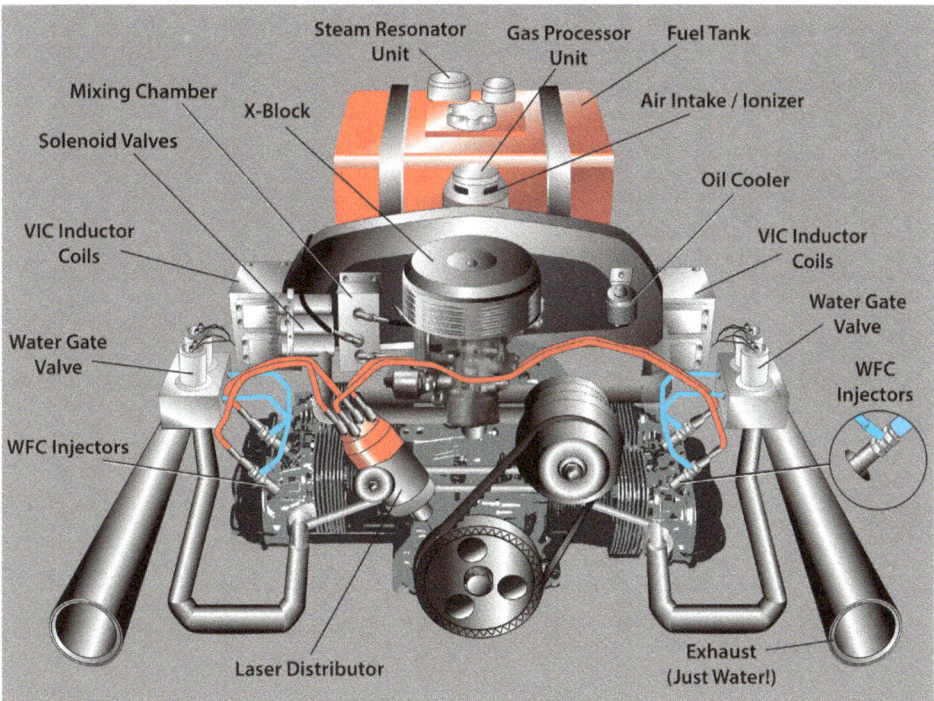

*The location of retrofit kit components on the 1600 CC Air-cooled VW Engine:
Illustration by the author using Adobe Illustrator.*

157

Unfortunately, since Stan's death, although this equipment has been salvaged, it has been found dismantled and most of the supporting paperwork has also gone missing (along with the safe that it was kept in) as yet nobody has been able to verify its functionality. However, after carefully studying all the patents, I have no reason to believe that it shouldn't work as intended. Stephen Meyer, Stan's surviving twin brother has seemingly chosen to distance himself from the work of Stan, perhaps scared off meeting the same fate, but significantly has developed a patent for Hydroxyl (hydrogen/oxygen) filling stations, based on the exact same principles! Certainly not what you'd expect if it was all a scam. I will look at that in detail in the next chapter.

Many parts of Stan Meyer's retrofit kit intended for any internal combustion engine, (whether it be fitted to a car, truck, train, bus, boat, or even a static generator!), had already been tried and tested on his earlier water fuelled car with some improvements, notably the simplification and miniaturisation of electronic components. I'll list the components with a detailed explanation of their function. The major difference is the water fuel injector, which incorporates water injection as well as an electrical voltage input. If you want to try to replicate it you can find the detailed drawings and 3D STL files for the parts on my website hydod.com. The electronic circuit designs are also available as well as completed boards from some sources (see Appendix 2).

So what other parts does it include and what purpose do they serve? Within the water tank is a 'Steam Resonator'. This is another very clever invention of Stan Meyer's which works in a similar way to a microwave oven. Its function is to heat the water to around 90 °C, so that when it enters the vacuum of the injectors it will immediately turn to vapour (note: not steam[1]), aiding the evaporation process and efficiency of the voltage-induced fracturing effect. This works by a process of 'particle oscillation' of the molecular dipoles (positive and negatively charged water molecules) the 'resonant action' of the water molecules, causing heat through friction during agitation and also helps to prevent icing up during cold weather. (Note: this technology can also be used for making hot water in a domestic environment.) Fitted alongside the water tank is the Gas Processor Unit which ionises the ambient air. (using infra-red LED light to ionise air molecules, which means to give them extra electrons. In effect, they will then have a 'negative' charge, because they've got electrons they can lose). A very clever part of the system (as mentioned in the last chapter), adds non-combustible gases directly from the exhaust system,

1. Water vapour is when the air contains water molecules, while steam is water heated until it turns into gas.

along with ionised ambient air, to the water. Since hydrogen burns too quickly, there is a necessity to retard the combustion by 'adjusting the burn rate' (to *"equal that of gasoline or fossil fuels, or even diesel fuels and gave us No. 1 retrofit capabilities"* he said). The 'non-combustible' gases slow down the burn rate as they are combined with the 'modulated and controlled speed' that the oxygen unites with the hydrogen, as it's the oxygen acting as an oxidant which in turn ignites the hydrogen. The three solenoid valves mix these three sources together in the correct proportions.

> *" Adding exhaust gases adjusts the burn rate to equal that of gasoline or fossil fuels, or even diesel fuel and gave us No.1 retrofit capabilities"*

> *"It runs very smoothly on water, and has an unusual sensation 'no lag! It's a fast responding fuel source, with a smooth operational performance".*

> *Stan Meyer*

Water from the tank is then fed into the injector through two Water Gate Valves on each side of the engine. The four Injectors simply replace the four regular spark plugs. These also have a high voltage electrical connection, converting the water as it arrives into what he calls 'an electronic polarisation' process, releasing what Stan called the "thermal explosive energy".

The Electronic Control Circuits located in miniature IC (Integrated Circuit) computer chips, allow 'regulation and control'. A 'Laser Accelerator' connected to the foot pedal and hooked up to the computer, converts the mechanical effort into electrical impulses. A new Laser Distributor, which fits between the conventional cap and distributor rotor assembly, also sends signals to the computer system, sensing the engine speed. The VIC (Voltage Increasing Coils) is a set of bifilar wound coils that boost the voltage considerably and provide the induced waveform that becomes the resonant frequency. It's the basic inductor/capacitor (LC) circuit that's at the heart of this technology. The alternating current, building potential in the coils, then charges up the magnetic field in the 'high power voltage zone' within the Injector. When this field collapses, as the current reverses again, it becomes an electric field again, tearing apart the water molecules and releasing the hydrogen ready for combustion. It also has an 'altitude adjustment' from sea level to mountain ranges, due to the way that air pressure affects the whole process.

Internal Combustion Engine

The Gas Combustion Stabilization Process (recycling non-combustible gases) is also applicable to operating an Internal Combustion Engine without changing Engine-Parts since the Gas Retarding Process allows the hydrogen "Bum-Rate" to "equal" the "Burn-Rate" of Gasoline or Diesel-Fuel. The engine provides its own non-combustible gases derived from Ambient Air undergoing the gas-combustion process. Engine temperature remains the same since The Gas Stabilization Process is used. The utilization and recycling of non-combustible gases, now, renders hydrogen gas as safe as Natural Gas or any other Fuel-Gas...allowing the Water Fuel Cell to become a Retrofit Energy System.

Memo WFC 423 DA & Memo WFC 421

(S. Meyer 'The birth of New Technology.pdf')

So what happened next? Did the death of Stan Meyer result in the complete stamping out of the whole water-fuel concept? (As the authorities and oil companies may have wished). Did his surviving twin brother disappear into oblivion? No. This story isn't about to end any day soon. Let's look at what happened to Stephen Meyer.

13

Stephen Meyer

Stanley Meyer's surviving twin brother

"Fuel cell and auto industries have been looking for methods and apparatus that can supply a source of hydrogen and oxygen for its new hybrid industry. This invention is such a device".

Stephen Meyer

The Meyer twins' surviving brother moved to Canada and didn't stop researching the idea of water fuel. In fact, he lodged patents for many things including a water fuelled hydroxy filling station to provide gas for the ever-increasing number of hydrogen-fuelled vehicles. He readdressed the situation and went way beyond what his brother had achieved in terms of efficiently releasing gas from water. He looked again at what was really going on and found out how to produce a "massive output" of hydrogen from water, with minimal energy input.

Stephen Meyer's "back to basics".

So is Stanley Meyer's 'Retrofit kit' the ultimate piece of technology in this whole water fuel saga? Well no, it's not exactly. Why? Because after his death his twin brother Stephen carried on where he had left off and in his own words *"went back to basics"*. He said in his radio interview with Jas Robey on blogtalkradio[1] in 2007 that he had done a reassessment to see *"what was really going on"* and as a result had gone *"elements of magnitude, way beyond what Stan had done"*, quite a claim. *"I went back to square one and did a lot of research over the last nine years to understand what is really happening. It's a magnitude over what Stan was doing"*.

Stephen Meyer's super-efficient output

For obvious reasons the one thing that Stephen doesn't do is build a water fuelled car (in case he endures the same fate as his twin brother). What he did though, is carry on with the exact same research and ended up achieving a patent[2] to produce what he calls 'hydroxyl gas' (which is an odd description for hydrogen/oxygen[3]) from a static filling point, intended to fill hydrogen-powered vehicles. It was a stroke of genius as it sidesteps the issue of producing a vehicle-mounted unit (or kit for the purpose) and anticipates the requirement to provide a way of refuelling the predicted glut of hydrogen fuel cell vehicles. In fact, an owner can install one in his own garage and use water to replenish it, using normal household mains power to produce hydrogen.

Now I see that as going back to Puharić's original research, which as previously mentioned I believe that Stan, being a Christian, didn't give Puharić credit due to his involvement in what he might consider 'black arts'. (Apologies to Stephen, Please correct me if I'm wrong. For the sake of humanity, the truth will come out in the end.) So why do I believe that and what evidence is there for it? Well, Stan's work usually centres around using a voltage input provided by an alternator (perfect for installations in vehicles, of course), or a simple oscillator circuit. However, Puharić, a qualified audiologist, used an audio oscillator and preamplifier and an audio modulator/power amplifier setup. A

1. Jas Robey BlogTalkRadio Stephen Meyer interview, https://bit.ly/3P5gTXH
2. Hydroxyl Filling Station: https://patents.google.com/patent/US20050246059A1/en
3. Stephen Meyer uses the term 'Hydroxyl' most probably because 'hydroxy' is trademarked. Also, a 'hydroxyl' is a covalently bonded atom consisting of oxygen and hydrogen -OH.

Motoring Journalist Quentin Willson refuelling the Hyundai ix35 Fuel cell car with Hydrogen Fuel at Nottingham University[1]

'resonance sensing resistor' ensures that the input is continually matched to the load. Low and behold, Stephen's electrolyser creates a "high-frequency ringing signal" creating a waveform almost identical to that of Puharić's patent[1]. 'Ringing' is an interesting choice of phrase as, for example when a bell rings, it has a fundamental note, but also comprises a range of harmonics. When a bell is 'tuned' a whole series of harmonics must be tuned with it. In the case of water fracturing, the cell must be tuned to the correct frequency in order for it to be 'resonant'. Puharić talked of the importance of the harmonics in his patent 'Method and Apparatus for Splitting Water Molecules[1]', an essential ingredient of 'ringing'. It is the ringing that enhances the production of 'hydroxyl' gas. Stephen also uses triple, not double, tubes, and his 'impedance matching circuit uses banks of 68uf capacitors, which when you use a capacitance/inductance calculator, gives a figure for the 'ringing' frequency required as 468.103Hz remarkably similar to that of Puharić's (bearing in mind that there will be variations due to other factors such as the quality of the water, distance between electrodes etc.).

1.   U.S. Patent 4,394,230 he filed in 1983 for a "Method and Apparatus for Splitting Water Molecules: https://bit.ly/3M9owtJ

In the financial world, traders look for an edge. It may be just a few per cent, but they add up so that they can profit from their transactions. In the quest for fuel efficiency, there are also many small changes that can add up to a significant gain. For example, the streamlining of the vehicle helps to reduce air friction and therefore drag, but also the weight, a lighter load requiring less energy to move it, increasing the tyre pressure for less friction, the removal of an unused roof rack to reduce drag further and most of all an economical driving style. In Stanley Meyer's patent, there are a few improvements that add up to make significantly more hydrogen. The biggest one is, of course, to use a resonant frequency, the main theme of this book. Once you hit that sweet spot, a huge increase in output matched with a lower input is observed, but also using a neutral as well as a cathode and anode, means that we have 18 'plates' (in this case, in the form of 'tubes') which means a lot more surface area producing gas. Also, since they are arranged with the outside tubes alternating between positive and negative, there is gas production *between* the clusters as well. Perhaps the main genius of this invention is the use of the dual three-phase input, which means that there is always maximum current potential as

Compared to a single-phase, two-wire system, a three-phase three-wire system transmits three times as much power for the same conductor size and voltage.

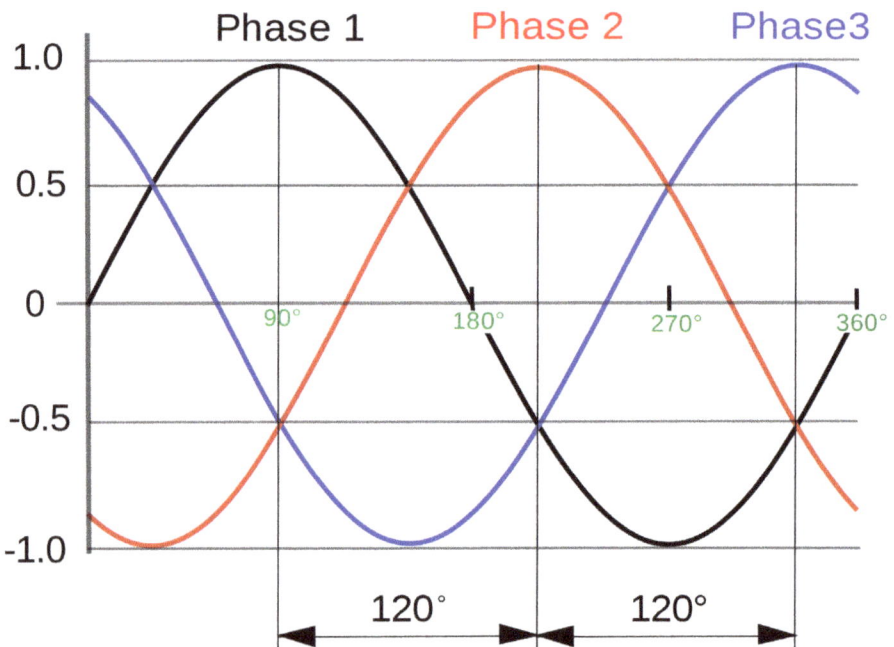

the wave cycles complement each other. (That's like triple the output x2!. See diagram on the next page) (Household energy is usually single-phase, about 50Hz here in the UK. Heavy industry, requiring significantly more power is usually three-phase). In Stephen's fuel cell there is always about 15v potential between the three tubes, even as the alternating frequency varies. The use of bifilar coils is also an energy gain. Winding the coils in the way Tesla describes[1], avoids opposing current flow, a kind of friction. When all the currents flow in the same direction around the coil, the output is boosted. This is the basis of Stephen and Stan Meyer's 'VIC' Voltage Intensifier Circuits. All this adds up to greater efficiency and a healthy hydrogen output, which means significantly less energy input and no laws of physics broken in the process.

"The invention is a computerized automatic on-site/ mobile hydroxyl gas-producing filling station that allows the products being produced to be used either by the hydrogen fuel cells installed in automobiles, trucks, buses, boats and land base generating applications or in any internal combustion engine".

Stephen Meyer

All this adds up to greater efficiency and a healthy hydrogen output, which means significantly less energy input and no laws of physics broken in the process!

1. Nikola Tesla's bifilar coil (United States patent 512,340 of 1894): https://patents.google.com/patent/US512340A/en?

Understanding Stephen Meyer's circuits

Firstly, Stephen Meyer's circuits bear all the hallmarks of both Puharić's and his brother Stan's work. In fact, I understand that Stephen was the electronics genius behind Stan's water-fuelled car. They are basically oscillating inductor/capacitor tank circuits as discussed throughout this book.

The patent consists of a dual, three-phase generator, six cards (circuit boards) and an array of six, triple-tubes (he calls 'wave-guides') in the water

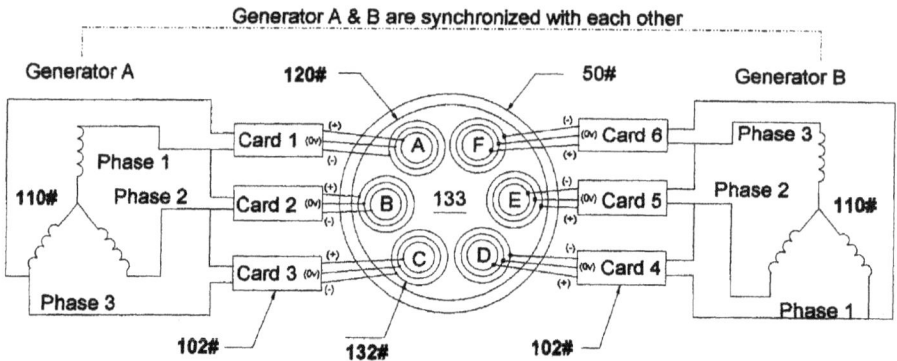

This diagram (Fig 3 in Patent USP Appln. 2005/0246059 – Hydroxyl Filling Station), shows a dual three-phase generator and the triple-tube array in its water bath (Water Fuel-Cell). The six cards (circuit boards) are detailed in the diagram on p172.

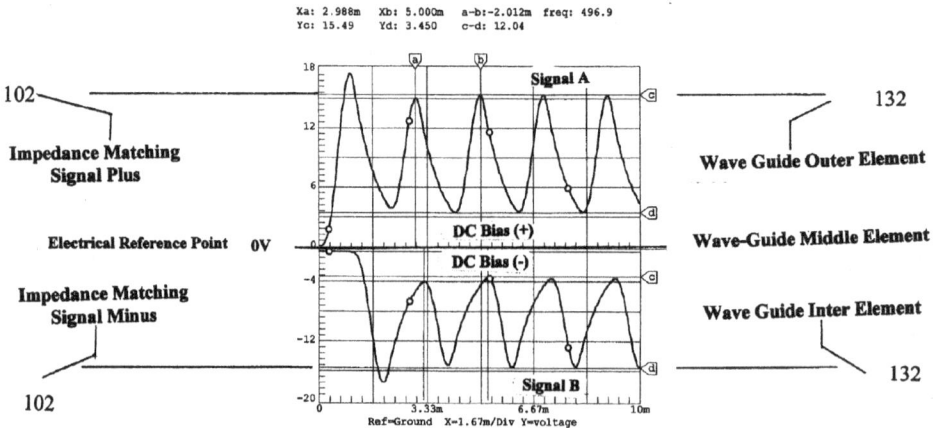

This diagram (Fig 5 in Patent USP Appln. 2005/0246059 – Hydroxyl Filling Station), shows the input signals with the voltage fluctuating between 15v, zero and -15v. Note the frequency of 496.9 Hz (cycles per second).

166

bath (water fuel-cell). There are also several coils, some with air cores and some with graphite cores, that also make them adjustable.

Firstly note that the water-fracturing carrier wave generated by these electronics is 496.9Hz (cycles per second), exactly the same as Puharić discovered. As well as being written on his 'Fig 5' (as shown on the previous page) is also evidenced by the 68μF capacitors (see explanation on the next page). Also to note that the switches sw1A, sw1B etc are in fact part of a relay to rapidly switch the current from the inner to the outer tubes. This would introduce the gates in the frequency pulse as required. It's now possible to use high-speed solid-state relays to achieve this. To this end, the circuit is also slightly incorrect. Perhaps it's an omission/error purposely introduced to make the design harder for competitors to replicate. (in which case I make no excuses for revealing this, as the world urgently needs this technology NOW).

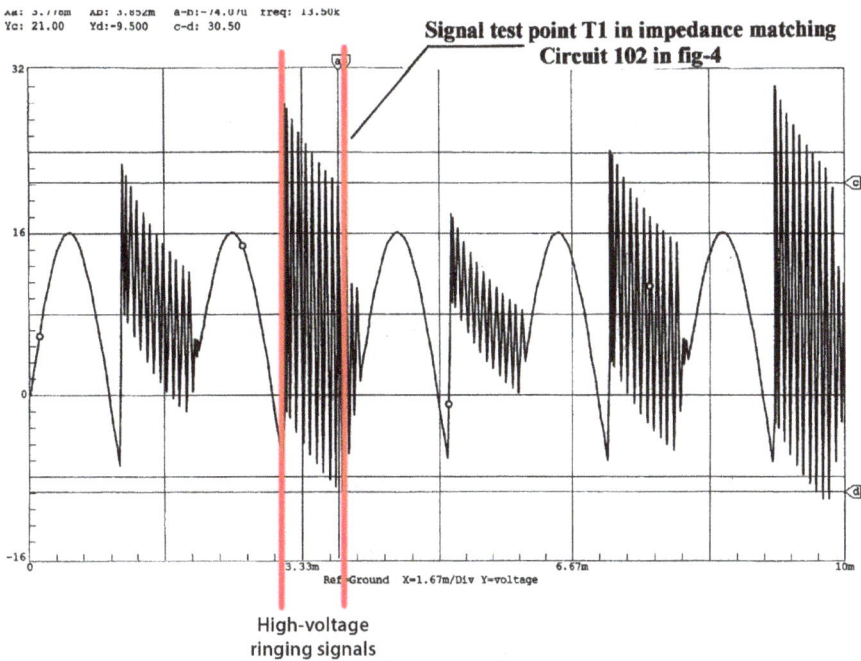

FIG-6

This diagram[1], shows the resulting water-fracturing resonant frequency between test points T1 and T2 in the impedance matching circuit[2], causing the high-frequency ringing signal that enhances the production of the hydroxyl gases.

1. Fig. 6 in Patent Hydroxyl Filling Station:
   https://patents.google.com/patent/US20050246059A1/en
2. 102 drawing Fig. 4 in Patent ~ Hydroxyl Filling Station.

The evolution of the Hydrogen

Method	Input	Equipment
School Electrolysis (Faraday)	+v — 1.23v to 15 Volts / Zero frequency!	**Direct Current (DC)** Hoffman (Tiny strip of platinum wire)
HHO 'Browns Gas' Hydroxy	12v — AC alternator 30 Amps — Alternating Current AC	**Alternating Current AC** Drycell 2 - 101 plates PWM Modulator
Puharić Patent: US4394230	Rippled Sawtooth — harmonics — Time	**Alternating Current AC** Array of 8 double alloytubes Fuel Cell
Stan Meyer 1600cc VW Car Patent:	12v AC alternator Gated pulse — 3 second gate (50% duty cycle) — 110v — 0v — 40A — Time	**Alternating Current AC**
Stan Meyer Retro-fit kit	12v — 12v AC alternator	**Alternating Current AC** water Fuel Injector
Stephen Meyer Hydroxl filling-stations	Dual 3-phase Generator — 15v — 0v — -15v — 3-phase	**Alternating Current AC** Array of 6 triple concentric stainless steel tubes

On-demand from Water Technology Stephen F. Meyer

Waveform	Notes
Zero frequency DC current! +v | 1.23v to 15 Volts Time	'Brute Force' Not efficient (Just a tiny stream of bubbles)
Gated pulse 3 second gate (50% duty cycle) 12v 0v Time	Max 2.5v per cell for max efficiency to prevent overheating
12v AC alternator (1 mAmp 10 volts - 1mA 26V)	Input is an amplitude modulated 90deg carrier sine wave (range: 200 to 100,000 Hz).
Resonant Frequency Pulse 12v Step Charge Time	Sufficient output to power a 1600cc engine
Resonant Frequency Pulse 12v Step Charge Time	Scalable to suit any engine size, cars and trucks
freq: 13.5k Efficient water-fracturing waveform	98% efficient Axial Flux, brushless generator "Massive production of hydrogen"

The circuits on the six cards, which is a basic L/C oscillator, include high-voltage 68μF electrolytic capacitors, these, along with the coils, produce the resonant frequency required. There is also an adjustable negative inductor, L2 with a ferrite core, which 'tunes' the circuit and accommodates to contaminants in the water so that the charge is always applied to the capacitor. The resonant frequency of 468.103Hz produces, as Stephen Meyer says *"high-frequency ringing signals (13.5 kHz) that contribute to the operation of the hydroxyl production"*. The voltage increase across the cell coincides with the production of the ringing signal (see fig 6, p167). A water-fracturing 'step-charge' effect results from a 'unipolar gated pulse', square waveform, charge/discharge cycle.

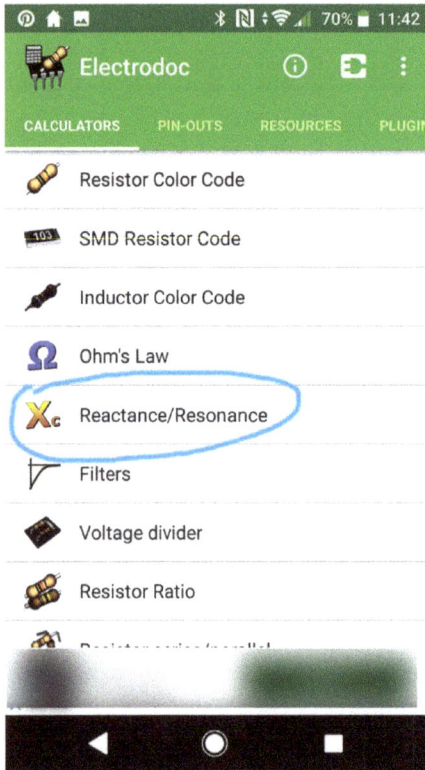

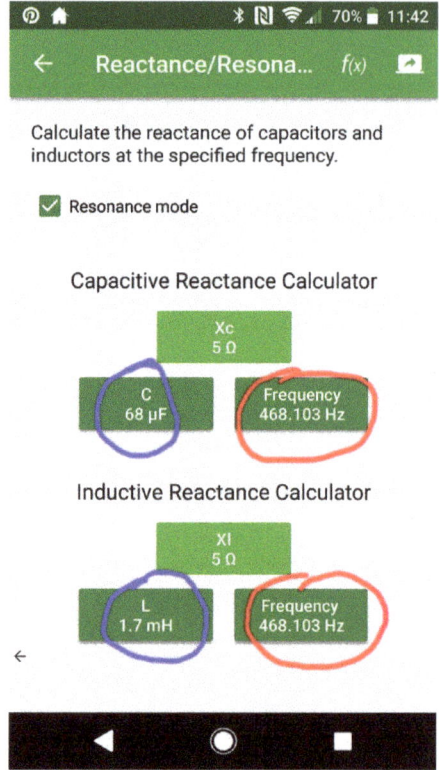

Using the component values in the circuit, let's calculate the 'resonant frequency'. (As shown, I downloaded the Android App 'Electrodoc' on my phone, but any other will do). Choosing the 'Reactance/Resonance' Calculator, put in the values for the inductors 1.7 mH and capacitors 68μF used in the circuit on the six cards (circled in blue). The result shows the frequency of resonance to be 468.103 (circled in red).

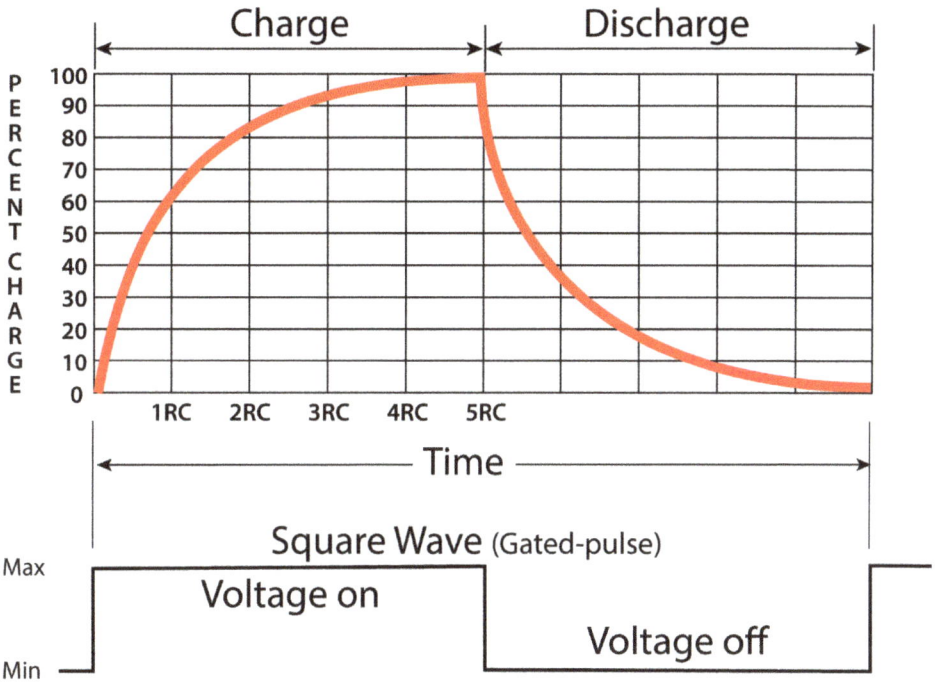

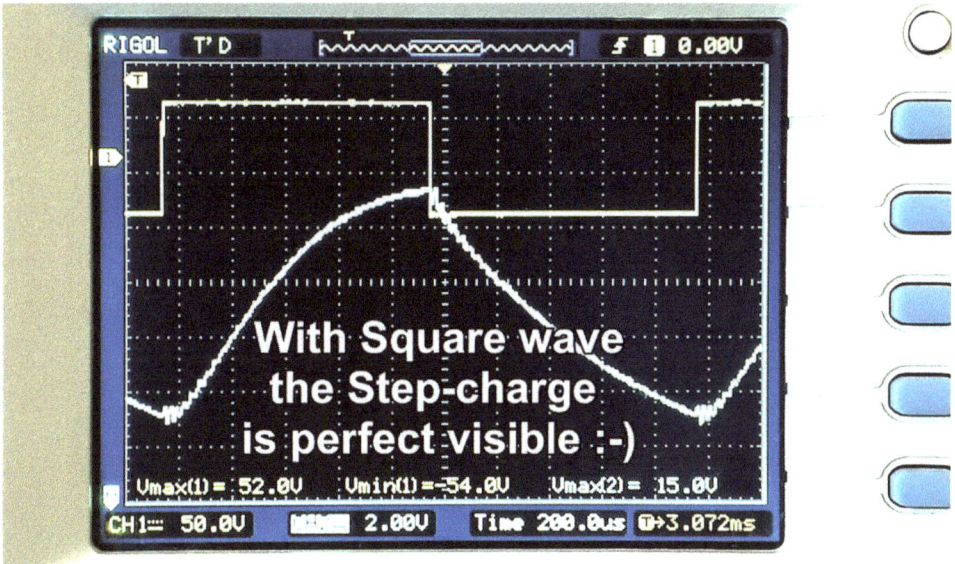

A screenshot of an oscilloscope by 'Petkov', a YouTuber (Valentin Petkov / valyonpz/), As an electronics engineer and researcher, who has thoroughly analysed Stephen Meyer's patent and uploaded his results[1]. It's the very latest video that he uploaded, about a month ago entitled 'Stephen Meyer`s 3-Electrode Water Fuel Cell'.

1. https://www.youtube.com/watch?v=vErg4LVz_b8 (See also chapter 17, page 223).

One of 6 Cards

Impedance Matching Circuit 102
Hydroxyl Filling Station

Isolation transformer
2 x 1000 turns, bifilar-wound, 0.6mm coated copper wire on an air core.

Slow-Blow Fuse 6A

1000V 6A Silicon Rectifiers

L1, L2 & L3 = 1.7mH Inductors.
(Bifilar wound, 380 turns of 0.6mm coated copper wire on air cores with 'adjustable' ferrite core on L3).

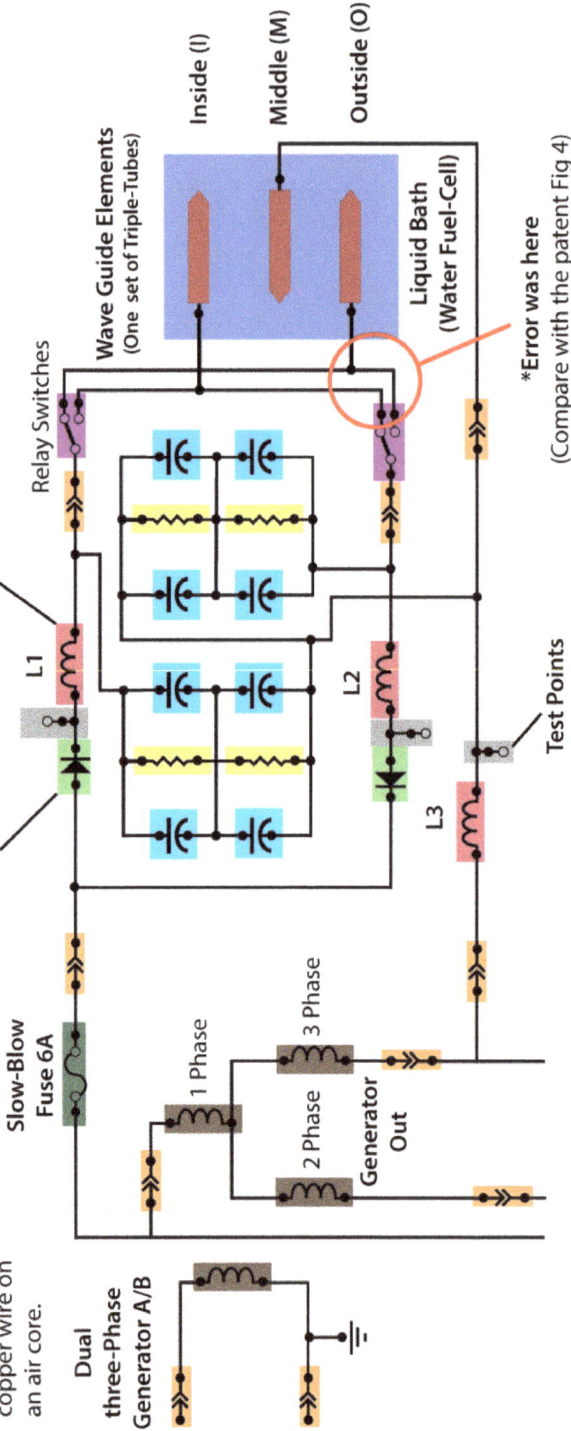

Relay Switches

Wave Guide Elements
(One set of Triple-Tubes)

Inside (I)

Middle (M)

Outside (O)

Liquid Bath (Water Fuel-Cell)

L1

L2

L3

Test Points

Dual three-Phase Generator A/B

Generator Out

1 Phase

2 Phase

3 Phase

*Error was here
(Compare with the patent Fig 4)

Test Points

Diodes (NTE5817)

Input Coil Inductors

'Wave Guides' (Tube Array)

Connectors

68µF 400v Capacitors

10 MΩ Resistors

1.7mH Inductors.

Relay Switches

Test Points

6A Slow-blow Fuse

Liquid Bath (Water Fuel-Cell)

Corrected Error: (Facilitates rapid switching between the inside and outside tubes).

Three-phase generator.

When we talk about using a 'resonant frequency' (The main theme of this book) and look at the waveform that a rotational body creates, that is a sine wave. An electric motor or generator produces such a wave. The electric power to our homes is usually a sine wave frequency of between 50-60 Hz (cycles per second). But that is 'single-phase'. A generator is capable of producing a dual-phase signal, or even a six-phase signal for every rotation. Let's look at that.

A six-phase motor, as it forms one revolution, can create six identical signals, all out of phase with each other. What that means, is that with each revolution we get six times the output compared to a single-phase generator. Also, crucially, the sum of all the phase current is zero and since the name of the game is efficiency, that's one way of achieving a low power input.

Some modern car alternators are six-phase in order to provide enough power for the ever growing number of electrically powered devices on board.

This is a dual, three-phase axial-flux generator. It is ideal for making wind turbines, for example, when you want to maximise the efficiency of rotation to produce electricity. This would be ideal for producing a current for water splitting. Some vehicle alternators are also 6-phase for the same reason.

XOGEN TECHNOLOGIES, Inc.

Now Xogen doesn't seem to have anything to do with vehicles or hydrogen production, but in fact, I believe that for a period of time Stephen Meyer worked with them to develop their fuel cells. What they say about the energy within water is very revealing as well as what energy is required to release it.

"If we examine one gallon of gasoline, the stored energy is huge. The power needed to release this energy is merely that of a spark, and as we all know, the energy contained in a spark is tiny".

If we examine one gallon of water, the energy stored is even higher than that of gasoline.

Xogen is the cutting-edge provider of Advanced Electro Oxidation technology to treat wastewater on-site to destroy ammonia and other contaminants.

"Question: Does Xogen's technology exploit stored energy in water? Answer: Yes...... It is often said that you cannot get more energy output than input. In its standard form, this is absolutely true. But let's take a closer look. If we examine one gallon of gasoline, the stored energy is huge. The power needed to release this energy is merely that of a spark, and as we all know, the energy contained in a spark is tiny. This example can be said by some to violate the laws of Physics. The energy of a spark input equals vast energy output. Therefore, it can be said that we violate nature's laws every time we turn the ignition key, but as we know, this is not true, as we are releasing the stored energy within the gasoline. The ignition system in a car engine is the energy release mechanism. If we examine one gallon of water, the energy stored is even higher than that of gasoline. Xogen doesn't release the stored energy with a spark, but with well-known scientific principles which have stood the test of time".

Xogen

https://xogen.ca/

Stanley's twin brother Stephen was the electrical wizard who helped his brother put together his world-beating car.

But he didn't want to die. So he decided not to make another car. He worked on trying to store hydrogen, which was a tricky idea because hydrogen leaks even through stainless steel.

He went back to the beginning to see what Puharić had done.

He completed all the scientific research, and multiplied the speed of production of hydrogen from water by using six-phase electricity.

Down in California "they" allowed him to produce high quality hydrogen and store it in petrol stations for anyone's use. Those storage tanks are still there.

But now we have almost the best combination of apparatus we can find.

We can get massive amounts of hydrogen easily from water by vibrating its resonant frequency. We can get huge amounts of it quickly by using multiphase alternating electrical voltage. We can do it simply right inside the engine, as we need it.

All we need to do now is for lots of scientific wizards to design simple and cheap conversion kits to install in petrol and diesel-guzzling vehicles, and mass produce those kits globally. What is good for one car will be good for *every internal combustion engine* in the whole world.

AND WE HAVE INSTANTLY SOLVED THE GLOBAL WARMING CRISIS.

I hope you can see the implications of Stephen Meyer's work. All the arguments about splitting water are about efficiency. Here is a solution hidden in plain sight with a simple but extremely efficient system for splitting water to release hydroxy gas. If you have a technical background I'm sure that you will have found his patent and electronic circuit interesting. It's a basic LC oscillator as has been the technical theme of this entire book.

14

The right idea but in the wrong time and place?

I can't help wondering whether these incredible innovations were discovered in the wrong time and place. It doesn't take a genius to realise that an idea which goes against the grain isn't going to get very far. Just look at the map I have outlined. It clearly shows the proximity of the pioneers of water fuel technology to the origins of the American oil industry. It's a tragedy that global warming and the energy crisis have got to the present extremes when a solution was under our noses all along. It's like stamping out dandelions thinking they're weeds when all along they had great health-giving properties. (The pharmaceutical giants probably won't like that comment either).

The wrong time and place?

Was this water-fuel technology doomed from the start, because of the time and place that it was developed right in the heartlands of the oil industry?

I'm including this map of part of North America to show the close proximity of many of the places significant to this research, particularly the key players of the Meyer twins and their predecessor Puharić.

Andrija Puharić did his thorough research into water splitting at Northwestern University, Evanston, Illinois. See how close that is to Oil City famous for the initial exploration and development of the petroleum industry? If you are going to develop an alternative to fossil fuel that's perhaps not the best place to do it. See also how close the birthplace and home of Andrija Puharić is to his acquaintance David Rockefeller in Cleveland, Ohio, who was the grandson of John D. Rockefeller Jr, founder of Standard Oil Co. He could either have supported him and propelled his ideas into mainstream use or killed them off permanently. He certainly had the means to do either. We know which he chose - the route of most personal gain at the expense of humanity and the planet. But then again, I suppose if your great family wealth is mainly from oil then it stands to reason. Sigh!

See how close Stanley Meyer who built the water-fuelled car is to Northwestern University. Did he get his idea of using a resonant frequency to split water molecules from God as he claimed? or from Andrija Puharić in Illinois just over 400 miles away? (I read somewhere that he was indeed a student of his for a while but unfortunately I can't find the reference to that now). People often discredit Stanley Meyer as a result of the fraud case where he was found to have committed "gross and egregious fraud", but would you expect him to get a fair trial in Ohio, home of Standard Oil Co.? I would be surprised if the courthouse wasn't absolutely filled with people whose lives were completely tied up in the local oil business.

As we have seen, Stanley Meyer died suddenly on March 20, 1998, while dining at a restaurant in Grove City, Franklin County. The Franklin County coroner ruled that Meyer died of a cerebral aneurysm. As I said previously, with plenty of oil money floating around the local population, a small bribe to a coroner would surely be no problem at all.

Pennsylvania oil rush

Titusville, Pennsylvania, is where the American oil-rush started[1]. Edwin L. Drake struck "rock oil" in the Oil Creek Valley in 1859 and the first oil wells were drilled there in 1861. This resulted in many oil wells and refineries springing up across the state. Due to the vast wealth resulting from it, there was a massive boom in oil with even more discovered from Texas to California. However, there is still a well-established oil industry in the area.

Standard Oil Company

The Standard Oil Co. was established in 1870 by John D. Rockefeller and Henry Flagler in Cleveland, Ohio, with refinery No.1 being built there in 1897. It became the largest oil refiner in the world and was so successful as the world's first and largest multinational corporation that it was deemed to be an illegal monopoly by the U.S. Supreme Court in 1911. As seen in chapter 10, Puharić knew David Rockefeller and his fate was in his hands.

Self-interested governments

Governments get a huge amount of revenue from fossil fuels and the American government in particular. They are also self-interested and if their military or other government bodies can benefit from a technology, they are quick to grab it for themselves and are often known to deem new technologies 'secret' and restrict their public exposure. This seems to be the case when an American Navy ship refuelled with seawater[3]. Both Puharić and Stanley Meyer are known to have been involved with government officials, Puharić was keen to stop them taking his patent and Stan actually having meetings with NASA in Cleveland[4] to discuss them using his patents.

---

1. Pennsylvania oil rush - Wikipedia: https://bit.ly/3w7CuWW
2. Standard Oil - Wikipedia: https://bit.ly/3LXMxoi
3. US Navy Successfully Converts Seawater Into Fuel: https://bit.ly/386TmFr
4. NASA's John H. Glenn Research Center at Lewis Field, Cleveland, Ohio. "Designs and develops innovative technology to advance NASA's missions in aeronautics and space exploration".

Map of the USA showing key

Northwestern University Evanston Illinois USA 1

NASA's John H. Glenn Research Center Lewis Field Cleveland, Ohio 6

Minnesota

Wisconsin

Andrija Puharić Chicago Illinois 3

Michigan

Ohio

Illinois

Indiana

Standard Oil Co. Cleveland Ohio 4

Kentucky

Tennessee

oma

places referred to in the text.

**Oil City
Venango County
Pennsylvania** 2

KEY

1. Northwestern University
(Research of Dr A. Puharić)

2. Oil City
(Start of American oil exploration)

3. Andrija Puharić
(Home of Dr A. Puharić)

4. Standard Oil Co.
(Founded by John D. Rockefeller Jr)

5. Stanley Meyer
(Inventor of the water-fuelled car)

6. NASA
(John H. Glenn Research Center)

New York

Pennsylvania

Virginia

North Carolina

**Stanley Meyer
Grove City
Franklin County
Ohio** 5

No Carbon Required

15

Pseudoscience?

Noun
A collection of beliefs or practices mistakenly regarded as being based on scientific methods

I'm hoping that this book will convince any sceptics and address any concerns that they have. In fact, I was a sceptic myself and went through a period where I, like many people, thought that releasing hydrogen from water efficiently wasn't possible, but having looked closer and investigated in-depth for a number of years, I now am firmly convinced.

This new water fuel technology has been attacked on several fronts. One way is to discredit the science behind it. In this new world of social media and online information, it's a powerful lobby. I hope that the details in this book will succeed in convincing mainstream science of the validity of its arguments.

There's nothing wrong with fringe science.

As long as it's *fringe* science and not *pseudoscience*. Science does not have all the answers. There are still plenty of opportunities for new and important discoveries. Even people like Nikola Tesla who gave us, among other things, AC electricity and radio, are still much misunderstood. Electricity and magnetism, as well as the properties of water itself, have many questions still to answer. Einstein thought that the three field equations of electricity, magnetism and gravity were so similar that they must have the same root cause (unified field theory), but he was wrong on this occasion, as gravity is entirely different from the other two. So to just pass something off as 'pseudoscience' without a second glance simply isn't an intelligent attitude to take.

It's not 'perpetual energy' (Anyone can edit Wikipedia)

I would also like to point out that there is a lot of misinformation on Wikipedia about this subject. They even have a page called 'List of pseudoscientific water fuel inventions[1]' "wherein common water is used to either augment or generate a fuel to power an engine, boiler or other source of power". (They only list two people, Stanley Meyer and Daniel Dingel, whom we shall look at in the next chapter). This *misinformation* is because it's possible for anyone to contribute or edit the information shown there. **It seems that certain parties are hell-bent on making sure that the idea of resonant-frequency water-splitting stays in the realms of pseudoscience.** I have corrected misleading sections myself, only to find that they have been changed back to their previous incorrect entries shortly afterwards. For example, included on the Wikipedia page, many sceptics are saying that Stanley Meyer's ideas amount to a type of perpetual energy machine[2]. This is incorrect as the gas released from the water is burnt up in the process, so it's depleted, being turned into mechanical energy, as in an internal combustion engine. Also, the process is not *making* the hydrogen, it is *releasing* it from the water and it's already highly combustible as is any hydrogen from any other source. In other words, we are not adding energy to make energy, we are adding energy to release the energy that is already contained within the gas. Does anyone say the same about oil? Does the refining process add energy to make it combustible? No! (Often people seem to forget that fossil fuels

1. List of pseudoscientific water fuel inventions: https://bit.ly/3KRrTox
2. Stanley Meyer's water fuel cell - Wikipedia: https://bit.ly/37wmGou

require a huge amount of energy to refine them, such as temperatures up to 400 degrees centigrade).

> # Water-fracturing devices are not *'perpetual motion machines'* because, in exactly the same way as fossil fuels, the water is used up in the process, albeit *much more slowly*.

What constitutes 'Scientific proof'?

In 2010, a company called "Future Energy Concepts Inc" purportedly succeeded in running a Dodge truck for 3000 miles on HHO gas from 12 gallons of water alone. There is a YouTube video to 'prove' it. However it's not really proof, and when I first saw it, I didn't believe it either. Why? And what has happened to them since? They say that the entire event was not covered by mainstream media. Why not? My guess is that no one else believed it either. "The technical event should have been the major news story of the last century!". The YouTube video shows the truck running and they 'show' that when switched to regular 'gas' the fuel quickly runs out. Hardly convincing 'proof' though. So, what is proof? What will convince everyone? The same thing occurs with Stanley Meyer's water car. The car had a switch on the dash, so that while testing, they could choose to make the car run on petrol (gas), or hydrogen on demand from water. So for a casual observer, how do they know which it is really running on? It could even be a dummy switch, right? In one video the two brothers are shown to be overjoyed that they have finally achieved success after years of work and the car is now running solely on water. Stanley points to the fuel tank that has been removed from the car altogether. But that is still NOT proof!

When I told someone recently that I believed it was possible to use water as fuel, he asked "*Well, where is your water-powered car then?*" The answer to that is simple enough. I didn't have the resources to make one and most likely couldn't raise the finance as I couldn't have convinced anyone to risk their capital in such a 'crackpot' scheme. (People often believe that if something like this is possible it will have become mainstream, but not with barriers like these). I also believe that a huge barrier is that scientists are concerned about their professional reputation and consign any ideas of water as a source of fuel to the same bin (trash can) as any ideas of 'flat earth' or 'UFO's.

When I was 8 years old at primary school, I did a complete project on the invention and development of the hovercraft in the 1950s. (A fitting precursor to my becoming a technical author!) Many of the world's greatest inventions and discoveries have been proved with deceptively simple experiments. Sir Christopher Cockerell did not have to build a hovercraft to prove that it was possible, he simply and effectively used an old domestic vacuum cleaner, a pair of kitchen scales and an empty coffee tin with a cat food tin inside it. This was sufficient to prove the concept. (But still didn't constitute 'scientific' proof). The vacuum cleaner switched to reverse, blowing downwards through the gap between the pair of concentric tins, with both ends open, produced a downwards thrust, which was then measured by the scales. Measurements were taken to show the relationship between the thrust of the electric motor and the power it was using, which he then compared to the thrust required to lift a particular weight off the ground, to determine if it could lift the source of the power. Personally convinced, he went on to demonstrate a working model, resulting in winning a contract with Saunders-Roe, a British aircraft and marine engineering company, to produce the SRN1 two-man hovercraft[1] (which I have seen first-hand as it is in a museum not far from my home town), for the National Research Development Corporation.

The hovercraft went on to prove itself with massive car-carrying versions regularly crossing the English Channel to France. It proved itself as an amphibious craft, being the only vehicle able to cross land, rivers and sea as well as marshy wetlands, impossible for any other type of craft. But it had inherent problems with directional control and high noise levels. The addition of a skirt to ride the unevenness of the terrain and prevent escaping thrust, particularly with waves, was a later addition that made a huge difference to its performance. (So inventions like Stanley Meyer's water car are not necessarily 'market ready' right out of the box and often require further development). During a considerable period in the history of the hovercraft, the UK Military, seeing the potential for warfare, took control of the patent and the invention and it was not seen by the public for many years as they kept it under wraps. (In a similar way, I believe that NASA and the US military are aware of the viability of hydrogen on demand from water as Stanley Meyer had meetings with them in the '80s. He talks about that in one of his videos[2]. They went on to produce fuel from water for some of their warships. (I personally know

1. "Saunders-Roe, Nautical 1" it was the first practical human-carrying hovercraft the SR.N1, which carried out several test programmes from 1959 to 1961
2. 'It runs on water' (The UK's Channel 4 documentary in 1980) - Stan Meyer: https://www.youtube.com/watch?v=t98UBY3GhhI

someone who lives in my neighbourhood that directly worked for the British military and NATO on hydrogen and oxygen production on submarines also using plain water as a source).

So, in order to scientifically prove a concept, you must use the correct scientific protocol. This is something that I have never seen done with hundreds of YouTube videos and claims relating to HHO and such technology. Perhaps often people are deliberately holding their cards close to their chests, in the hope of making a personal fortune from their discoveries, and instead disappear into oblivion. But sometimes I think they just don't know how to use the correct scientific process. They make shaky videos, often out of focus and poorly lit, without proper explanation. Their 'proof' of hydrogen production is to light the bubbles to make them 'pop', but the question of how much hydrogen has been produced with what amount of current and at what voltage remains unanswered. I intend to address this issue by preparing a proper scientific report for submission to a highly respected scientific journal for peer review. That is something I don't believe anyone has done so far. (And it's not cheap, running into thousands of pounds). Yes, patents exist, as I have shown, but until this is done it simply won't be accepted by the scientific community.

What is the scientific method?

There are six steps to a proper scientific method which include:

1. Ask a question about something you have observed.
2. Do background research to learn what knowledge exists on the subject.
3. Construct a hypothesis.
4. Experiment to test the hypothesis.
5. Analysing data from your experiment to draw conclusions.
6. Publish results for peer review.

For example, 1. Is 'resonant frequency electrolysis' more efficient at producing hydrogen? 2. Find out if anyone else has published anything on the subject. 3. Propose that if a greater quantity of gas is produced consistently, by using a resonant frequency rather than, (as Faraday did), a DC current, then it must be more efficient. 4. Test different frequencies and straight DC currents over a period of time, using the exact same calibrated equipment and measuring the quantity of gas produced over time. 5. Compare results. 6. Publish results in a reputable scientific journal for peer review (who may spot errors, or notice incorrect assumptions in your methods or calculations).

Taking measurements

So, how can you measure hydrogen production? Firstly accurate instruments that have been properly calibrated should be used. Measuring voltage and current is very easy, but measuring a quantity or flow of gas is a little more tricky. I have researched this and came across the 'displacement method'. That is where the gas is collected in a vessel that contains water and pushes the water out as the hydrogen accumulates over a set period of time. This water can then be collected and weighed. The volume of water expelled will be the same as the volume of gas that is produced and therefore accurate measurements can be made.

I have not seen much evidence of YouTubers using the correct scientific process, in particular using measurement data to assess the success of a particular setup, in order to advance the process to the next stage. (That's not to say they are definitely not doing so, of course).

Does it break the laws of physics?

People often claim that the idea of splitting water to release hydrogen and oxygen breaks the laws of physics and they use that as an excuse to discredit anyone attempting to do so. It does not break any natural laws if you really understand what is going on.

Rewriting the law of thermodynamics

If you look at the history of the Law of Thermodynamics you'll find a lot of confusion. It is comprised of several people's research combined. Disagreements and additions are still ongoing, as is the case with most theories. They stand until someone comes along to disprove or at least throw new light upon it. In fact, there were three laws and now a fourth has been added called the 'zeroth'!

It was penned during the age of steam. It was essentially an observation of the difference between a coal fire and the atmosphere. People who are still living in the past quote it whilst still convinced that electrolysis is so inefficient that the amount of energy going into a chemical process is about equal to the output. That may have been true with steam, but certainly not with our modern technology and all its refinements. We can achieve much greater efficiency using modern technology and materials! Babbage, the inventor of the first computing machine, would be amazed that I can sit in my favourite chair whilst communicating with someone through a wireless phone, or using the Internet to order anything from Amazon for next-day delivery. (In fact a package has just arrived with some parts that I had ordered from China while I'm typing this). He would probably not believe that I can design something on a home computer screen and after uploading the design wirelessly, have a part 3D printed or precision cut with a laser, or a one-off or mass-produced circuit made and delivered to my door. He would be incredulous that a palm-sized computer had so much power to compute. Stephenson would be similarly amazed that we have cars (but not the model T Ford until 1913 - another 80 years) boats, even aeroplanes and space rockets weighing hundred of tons. (He would still have to wait for another 70 years until 1903 before the first plane), all powered by all manner of propulsion systems. In 1872, American George Brayton invented the first commercial liquid-fueled internal combustion engine. Steam power is mostly now resigned to museums and tourist attractions, except for a few coal-fired power station generators, although many are now gas turbine. So maybe it's about time to update the old, out-of-date Law of Thermodynamics. If the process is more efficient it means you'll unlock energy that already exists within the water molecule.

Why people don't succeed

Of course, one reason is that they may not be doing it correctly!

One YouTuber exclaimed that nobody succeeds in getting Stan Meyer's resonant cell to work because you need a huge stack of expensive equipment, worth thousands of dollars. Why? Because there are so many variables, so to find all the correct settings simultaneously was a near impossibility. Another YouTuber noted that the type of water changed the resistance values and just holding your hand near the WFC changed the capacitance. (Similar to those lamps that you touch to turn on, or indeed a 'touch screen).

Also as we've seen in an earlier chapter, during the process of electrolysis within the WFC, bubbles of gas build up on the surface of the cell. These bubbles have the effect of turning the cell into a 'double layer' capacitor, reducing the capacitance or stopping the process altogether. Puharić noted that 'tapping' the cell released them, resulting in the process continuing.

Stanley Meyer had the genius idea of mixing the hydrogen with a little non-combustible exhaust gas to retard its combustion a little, as it was exploding too quickly and burning too hot on its own. This meant that it would match the burn rate of conventional fossil fuels.

YouTubers' videos are mostly very amateurish, out of focus, disorganised (like their workbenches), and jumping around with little explanation of what is happening. And no one ever, ever measures the output of hydrogen! If you don't measure the output of hydrogen for a given input, it proves nothing!

Some people talk of the need to prepare the surfaces of the stainless steel as a way of 'conditioning' it (it produces an oxidised surface - a white powder coating) (Chromium oxide dielectric). Some also say insulation is required on the outsides of outer tubes and inside the inner tubes, which may help.

Working in the dark.

Science has so many questions still to answer and particularly so with the technology of fracturing water molecules to release hydrogen. One reason for that is partly because what we are working with is invisible. Electricity is invisible, magnetism is invisible, gas is invisible and the hydrogen flame is so pure that even that is hardly visible in daylight. This makes this technology hard to control as precisely as is required to get good, consistent results. For example, how do you know if a voltage is leaking around, rather than passing through a metal plate that forms part of a Fuel cell? Exacting tolerances are crucial for success, perhaps even more so than for traditional fossil fuel-powered technologies. Magnetism, a necessary part of the process, still holds mysteries for science. One way of demonstrating a magnetic field is to use iron filings on a board with the magnet fixed below. However, that only shows a two-dimensional representation. New theories believe that the magnetic field is actually a vortex. Another similar theory is that the lines of force are always complete circles and so in fact also go around the corners of the magnet. What is known about static electricity? It is often considered just a nuisance causing eddy currents and spurious spikes in circuits as well as radio interference. Tesla did a lot of experiments with that, convinced that it would provide endless 'free' Energy, but little seems to have been done along similar lines since. The many differing methods outlined in this book are often seen to be utilising alternating collapsing electric and magnetic fields to break the molecular bonds of the water (each one dumping its charge to the other). Tesla bifilar coils also seem to be largely overlooked, but essential to the success of the water-fuel process as they massively increase the power output with little current. Many of the references to these aspects lead sceptics to deem the outlined technologies 'pseudoscience' and the perpetrators 'conmen' or worse. Sadly, that often means that the valuable insights of these cutting-edge innovators are ignored, passed by and forgotten, when, in fact, they are revolutionary and in a time of global climate catastrophe, very likely essential to our survival.

Convincing the sceptics

I'm hoping that this book will convince any sceptics and address any concerns that they have. In fact, I was a sceptic myself and went through a period where I thought it wasn't possible, but having looked closer and investigated in-depth for a number of years, I now am firmly convinced.

There are going to be many people connected with the oil industry and people who have heavily invested in it, who hope that all this is baloney.

I urge scientists and technical people to look again. It may at first seem an incredible feat to perform and yet, if we can all work on this together the benefits are enormous. Collaboration is the key to success.

So what about the proof?

At the beginning of the book we saw the proof in the 'Journal of Plasma Physics' (Volume 63 Issue 2), where the 'Arc-liberated chemical energy exceeds electrical input' I'm going to take another look at that in the next chapter on Plasma).

1. Arc-liberated chemical energy exceeds electrical input energy: https://bit.ly/3KRaXhH

So, new science can seem at first to be pseudoscience, until you look a lot closer. I urge you to look as closely as you can to this water-splitting technology as I have done, with an open mind.

All truth passes through three stages:

First, it is ridiculed;
Second, it is violently opposed;
Third, it is accepted as self-evident.

Arthur Schopenhauer (1788-1860)

One area of science that needs a lot more study is plasma electrolysis. To get plasma we need very high temperatures, but we can do that with a momentary high voltage spark. Plasma electrolysis has been shown to release the explosive energy contained within water with very little energy input. Let's look at that a bit more closely.

16

Plasma

"Plasma is the most extreme state of matter in the universe, sometimes called "the fourth state of matter."

Dr Melanie Windridge, Tokamak Energy.

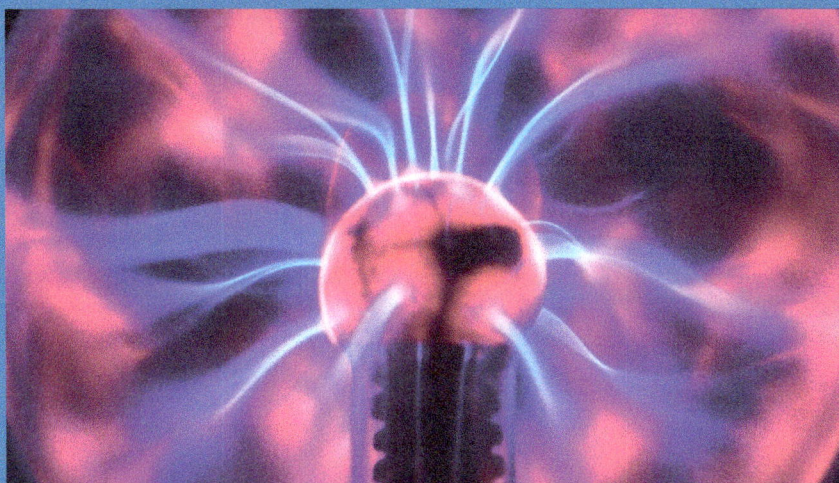

Detail of the plasma sparks within a Tesla coil.

This is a fascinating science and I find that the more I look the more interesting it gets. One thing that I have come across is the use of a plasma spark to release hydrogen and oxygen from water. It's early days and needs much more research, but from what I can see, plasma holds a lot of promise. The great Nichola Tesla was trying to tell us something.

Plasma

When very strong heat is applied, certain substances undergo another change in state to plasma as the electrons are stripped from their respective atoms, creating free electrons and positive ions. Although there are both positive and negative particles, plasma is neutral overall, as there are equal amounts of oppositely charged particles. Since there are free electrons present, substances in a plasma form can conduct electricity. Gas cannot conduct electricity, but plasma can. Naturally occurring plasma includes lightning and the Northern lights and stars, including our sun, which are all just balls of hot plasma. Plasma can be found in neon light bulbs and neon signs. When an electric current is passed through the Mercury vapour in a fluorescent tube, it heats up the gases sufficiently to strip the electrons and create plasma. Plasma TVs are made possible due to the state of plasma. The screen is made of thousands of fluorescent light electrode pixels emitting RGB colours, the combination of these colours producing any possible colour.

The sun

The sun is a huge ball of plasma due to the extremely high temperatures of fusion during the process of hydrogen turning to helium. It's the sun's massive gravitational pull that compresses the gas and causes the fusion. The sun's magnetic field also causes electromagnetic flares to escape from the sunspots on its surface.

Tesla coil

A partial vacuum within the Tesla sphere contains electrons.

Although it takes a great amount of heat to produce plasma, we can achieve that without much energy input by boosting the voltage. That's an extremely high voltage, however, perhaps in the millions. If that sounds a bit far-fetched, we do in fact already do this in our cars, not in the millions, but in the tens of thousands. A car coil typically boosts the voltage from a mere 12 volts to a huge 20,000 volts or more. It's this that jumps across the precise spark gap to ignite the fuel/air mixture in the cylinders. Instead of just igniting fuel, we can also use this method to split water molecules into useful gas. I said that it takes very high heat to produce plasma, but we can create the heat for the plasma by getting the extremely high voltage to jump the spark gap. This is one of the things that the great Nikola Tesla was researching of course, as can be witnessed in his Tesla sphere that everyone has probably come across. It's usually only seen as a novelty though, but wait, isn't it free energy? (In this case, 'free' energy means energy that is from the environment that we can harness). That's obviously what interested Tesla.

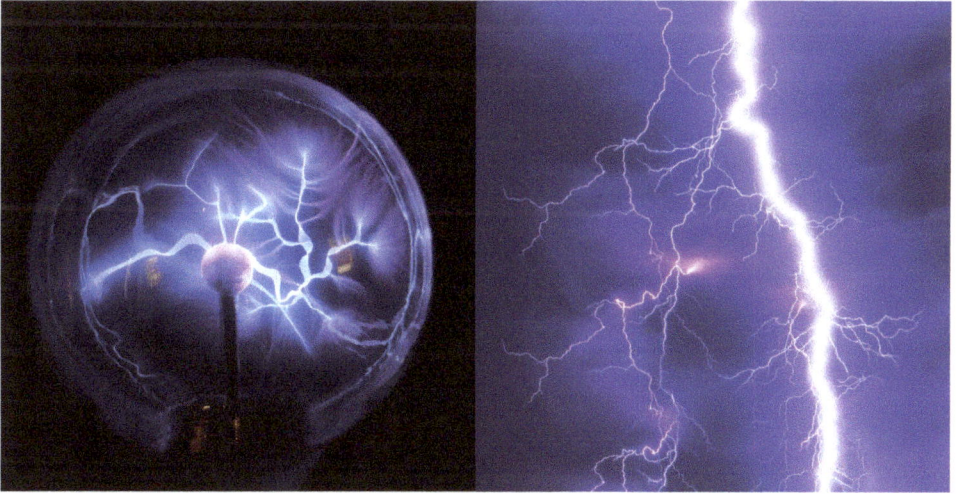

Very high voltages create plasma within a Tesla coil (Left) and in the atmosphere as lightning (right).

A typical automotive ignition coil boosts the vehicle's typical 12v to 20,000v or more for the ignition spark

A solar flare of electromagnetic radiation on the surface of the sun,

A confession

I must confess that this is a fairly new area for me and much more research is required on my part. The initial findings are very encouraging though, especially with the scientific paper concluding that much less energy input was required than obtained by the output when fracturing water using a plasma spark. I have also seen evidence on YouTube of people who were producing a massive amount of oxyhydrogen using a plasma spark but unfortunately, they stopped posting videos four years ago and they didn't explain exactly how they were doing it. It's also very dangerous. The report shown below, talks about an explosion that occurred at Hokkaido University when experimenting with plasma electrolysis, despite them only using a low voltage and current, which is interesting, to say the least!

Extreme plasma voltages

Warning

Be very careful, plasma electrolysis is highly dangerous,

even at very low voltage and current!

(Tip: Never use glass, even strengthened

glass like Pyrex for electrolysis)

On January 24, 2005, at around 4:00 p.m. an explosion occurred during a plasma electrolysis experiment at the Division of Quantum Energy Engineering Hokkaido University[1].

*"Soon after ordinary electrolysis began, the voltage was increased to **20 V** and current to **1.5 A**. 5 or 6 seconds later, a bright white flash was seen on the lower portion of the cathode. The light expanded and at the same instant the cell exploded".*

---

1.   Accident Report: https://bit.ly/3x6pH8T

Extreme Plasma voltages can help to create water plasma. Perhaps you learned at school that everything has three states of matter, solid, liquid and gas? Well, plasma is the fourth state. It's something like an energised gas. Water in that state is easier to fracture and since it is already ionised it helps make the process more efficient.

Plasma electrolysis experiment proves that you can get more energy out of the process than you put in.

Journal of Plasma Physics: Volume 63 Issue 2

'Arc-liberated chemical energy exceeds electrical input energy[1]'

(Published online by Cambridge University Press: 01 February 2000

by PETER GRANEAU, NEAL GRANEAU & GEORGE HATHAWAY)

Extract

"This scientific paper reports the first experimental results in which the kinetic energy of cold fog, generated in a water arc plasma, exceeds the electrical energy supplied to form and maintain the arc. The cold fog explosion is produced by breaking down a small quantity of liquid water and passing a kiloampere current pulse through the plasma. The 90-year history of unusually strong water arc explosions is reviewed. Experimental observations leave little doubt that internal water energy is being liberated by the sudden electro-dynamic conversion of about one-third of the water to dense fog. High-speed photography reveals that the fog expels itself from the water at supersonic velocities. The loss of intermolecular bond energy in the conversion from liquid to fog must be the source of the explosion energy".

1. 'Arc-liberated chemical energy exceeds electrical input energy':
   https://bit.ly/3KRaXhH

Nexus plasma spark generator

There's a patented invention called the 'Nexus Plasma Spark Generator', which claims improvements in engine performance for a very low cost. It is based on a 12-volt to 110-volt inverter powering a microwave transformer which produces a very high voltage, to create the spark in the engine, but with a current draw of only 100 milliamps at 70 mph. The plasma spark ignites a water mist spray fed into the air intake. A YouTube video shows the device in operation and the website has a good explanation and a circuit diagram detailing the exact operation. The inventor, Bill Cozzolino, demonstrates the device working in his VW air-cooled bus as an addition to the regular fuel supply. He has uncoupled one spark plug to demonstrate the very noticeable difference in the spark when the device is on and off. He claims to have had the engine running on water alone, but as he says *"It does not run well, it coughs, splutters, misses, farts and moans, but it does fire on water alone!"*

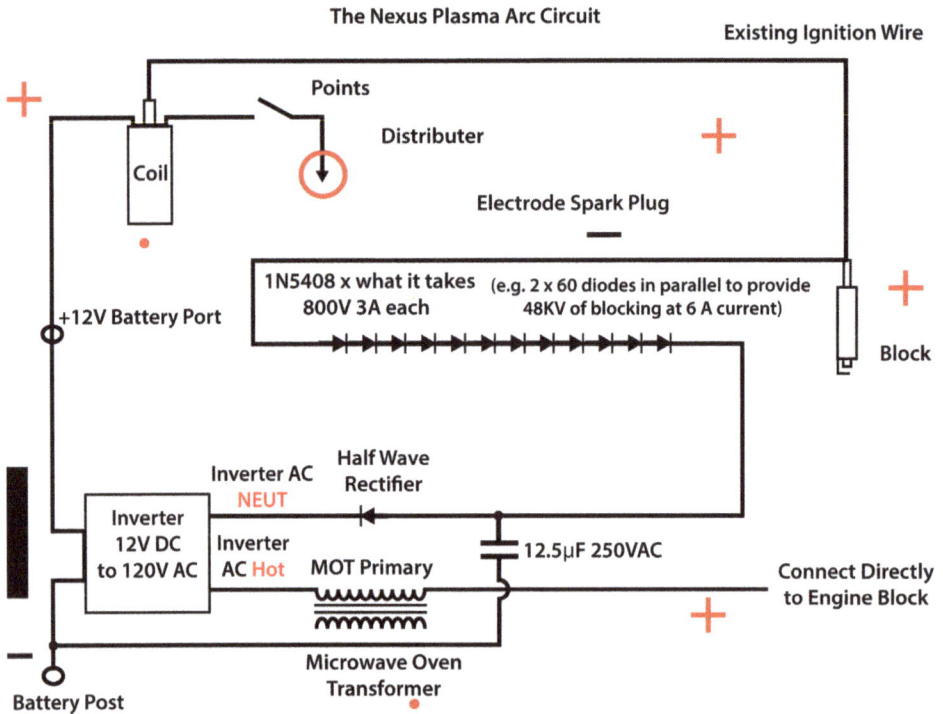

The Nexus Plasma Arc Circuit

Existing Ignition Wire

Points

Distributer

Coil

Electrode Spark Plug

1N5408 x what it takes (e.g. 2 x 60 diodes in parallel to provide
800V 3A each 48KV of blocking at 6 A current)

+12V Battery Port

Block

Half Wave
Rectifier

Inverter AC NEUT

Inverter
12V DC
to 120V AC

Inverter
AC Hot MOT Primary

12.5μF 250VAC

Connect Directly
to Engine Block

Microwave Oven
Transformer

Battery Post

Bill Cozzolino's automotive circuit for producing a plasma spark. Achieving a very high voltage using an extremely low current, is used to ignite vaporised water in an air-cooled VW engine. He got a VW Bus running on just water. The unconventional way that it's wired is intentional. It's the secret to its operation, using opposing currents.

According to Bill, it's just a matter of more research to further refine the control of the spark. He said during a radio interview, that it was important to note the difference between a plasma arc and a spark across a spark gap. The thing that did initially surprise me were the long strings of series diodes that he is using. He explains that the reason for this is to prevent 'crossfire' of the high voltage between the four spark plugs or to ground (the engine block) as, similar to lightning, the path to ground is the easiest route to earth, the current would otherwise completely bypass the spark plugs. That's through anything close at hand, even you! In other words, they can be quite dangerous. That's the reason for lightning conductors, for example. However, after a little research, I found that it is possible to buy 'Voltage blocking diode boards' for just such purposes, with banks of diodes, typically rated at 60 amps and 1200 peak inverse voltage, providing a 'blocking voltage of around 7200volts!

Nine blocking diodes on a circuit board in series. This one is rated to handle an incredible 16200V, with each diode rated for 60 Amps and 1800 peak inverse voltage.

Bill Cozzolino's website Skyhero.com has unfortunately disappeared, but using the 'Wayback Machine[1]", you can still see parts of it and download the pdf and instructions for the 'Nexus Plasma Arc Circuit[2,]' (on the previous page).

However, there's an excellent publication all about plasma spark electrolysis. 'WaterSparkPlug.pdf[3]' from the 'Panacea-BOCAF' (stands for 'Building Our Children A Future').

1. Internet Archive Waybackmachine: https://web.archive.org/web/
2. Nexus Plasma Arc Circuit: https://bit.ly/3MW324A
3. WaterSparkPlug.pdf: http://www.panaceatech.org/Water%20Spark%20Plug.pdf

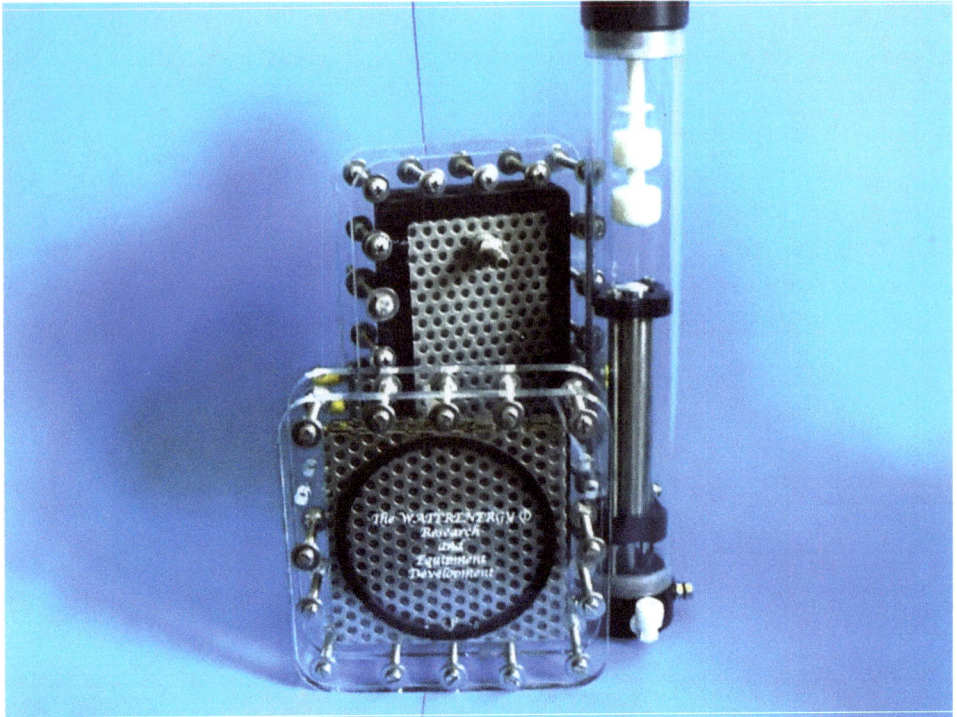

Drycells and other equipment on offer by 'The WaterEnergy1' Team.

Detail of the plasma flame showing through the perforated stainless steel plate.

High voltage HHO, The R.E.M.F. high voltage plasma energy,

You can of course get a spark to work underwater and this seems to be what these people are doing. YouTube videos posted on 'THEWATERENERGY1' channel, show quite clearly that they are using a high voltage spark to achieve a massive output of hydrogen. In fact, although they do not disclose exactly what the output is, it's obvious by visual observations that it's way more than any other form of electrolysis. They claim it's about 300 LPM and I can quite believe that looking at the videos. This obviously needs further research. Another interesting thing is that they are using perforated stainless steel plates. That is one way of hugely increasing the plate area. As I have said before, it's by increasing plate area that we can boost gas output.

"..at a plasma state, our power input drops almost to zero amps".

'THEWATERENERGY1'

Editor's note: These people are Dutch, consequently I decided not to make any edits to their text for fear of misinterpreting anything. Make of it what you will.

REMF Plasma: resonant excitation of a moment field (REMF)

"We are able to reverse the process, plasma state at a certain level/moment is no longer power input but power output,...(It) Means in this process (we) generated Hydrogen Oxygen Heat and useful electricity".

*"Starting the processes; Is to bring your cell **at a certain temperature and bring the cell at a certain pressure**, As soon we reach our desired pressure, the processes continue with oscillations of the cell pressure, where our so desire effects arise, continuing those proceeding **will bring our cell in resonance**, in combination with surrounding R.E.M.Fields in the cavities, Vortexes appears at the service of the holes of the plate, in case of tubes between the tubes, Editing of the differential pressure which, intentionally ascent will be held in the cell, we have as result that all the little bubbles turn into mega bubbles, passing through apply R.E.M fields, Gaining more and more Energy, by absorbing apply photon energy of several energy spectrum/frequency, As well known by all physics, Each energy level have their own energy spectrum in form of light frequency from infrared down to the X Rays,(More update texts will follow)"*

'THEWATERENERGY1'

201

"1. Device arranged for the production of H_2 and/or O_2 gas and/or electric and/or thermal energy;

The device comprising of a container containing a quantity of water, at least two electrodes extending in the container and connected to connection terminals outside the container, and an H_2 outlet and an O_2 outlet, both extending from the inside to the outside of the container; the electrodes consisting of metal tubes, rods, or plates, of any form, which are perforated.

The device, moreover, comprises of an electric power supply arranged to supply electric power to said electrodes via an electric circuit,

said power supply having as input either DC or AC and having the possibility to supply electric power (DC or AC) at varying frequencies of any form, with a possibility to fix the frequency; the frequency needs to be fixed when the right frequency is found when the special effect occurs that the quantity of water in the container is very rapidly converted into H_2 and O_2. Varying back the frequency can limit this process and control the quantities produced. Limiting the quantity of water in the container also limits the maximum quantity of gas produced."

THEWATERENERGY1

Spark modules

Another person who has successfully used plasma sparks to fracture water, goes by the YouTube name of Ritalie (aka 'Charles'). He sells circuits, eBooks and parts. His circuits use standard car spark modules, or microwave transformers can also be used to massively boost voltages. His circuits can be used for battery charging ('Radiant Battery Charger'). His circuits for water-splitting are also designed to automatically lock on to the correct frequency, because yes, although this is about plasma, it's still a resonant frequency that we are talking about, albeit a much higher one. 'Ritalie' also seems to have harnessed 'back EMF' (electromotive force) a normally wasted backflow current that is usually considered a nuisance. (He calls it 'radiant energy' and has even been able

1. https://www.ritalie.com/ and https://ritalie.com/store/

to use it to charge batteries (using an "auto-tune" self-resonating, solid-state circuit). Indeed in electronics, resonance is often considered to be a nuisance and to be avoided as it causes unwanted eddy currents and radio interference. In cars, for example, this radio interference from the engine's sparks needs to be suppressed using a capacitor to filter out the frequency in order to eliminate distortion, humming and hissing over audio systems.

Lightning

It's a killer! Everyone knows that lightning strikes can kill and even relatively low voltages as those used in our domestic appliances can too if care is not taken. In the UK we use around 240v AC, and Europe and the USA around 110v AC. The AC means that the current is passing back and forth to do the work. (Think of the ocean waves and the power that they have as they ebb and flow). Electrons do not travel along a wire as such, they just jump a tiny distance from atom to atom. It's the combined effect of all those, in a chain reaction that creates the apparent 'flow' of current. It's the same with waves in water. Although the waves are moving, the actual water particles are mainly just moving up and down, (much like the 'Mexican' wave at football matches.) It was Tesla that introduced the AC system in a fierce competition with Edison. (There was even a recent film (movie) about that called 'The Current Wars'). Tesla won out because AC suffers fewer losses of power over long distances when transmitted over cables across the countryside. In fact, the power is increased to voltages as high as 500,000v, which is more efficient to transmit across many miles of cables and then reduced again in local 'sub-stations' to much lower domestic voltages.

Tesla experimented with extreme voltages in his quest to find an unlimited supply of energy provided by the magnetic and electric fields of the planet itself.

Benjamin Franklin (October 18, 1785 – November 5, 1788), one of the Founding Fathers of the United States, invented the lightning rod, which when mounted on the highest point of a building, conducts the very high voltages generated safely to earth (ground). He also famously experimented with a kite flown in an electrical storm, in order to prove that it was electricity. (Luckily he was standing on an insulator!). (It was for a similar reason that the famous Hindenburg airship ignited, see chapter 3).

Microwave transformers

In order to create microwave energy to cook with, microwave ovens boost the voltage using a very powerful transformer. These transformers can be repurposed to create water plasma, which can then more easily be split into its respective gases.

> *"There IS no energy shortage; only grey matter."*
> *Donald Lee Smith*

Donald Lee Smith, (1928 - 2010)

It's another case of the voltage doing the work, rather than amperage.

I have to mention the work of the American Don Smith. Why? Because he did some very important and interesting work on resonant frequency plasma 'free energy' or 'over unity' machines[1] (something that generates more energy than it uses). This can also be used for super-efficient water electrolysis.

His patents were evidently so good that they were suppressed in a typical way. They were bought and promptly hidden from the public so that it was illegal to build one of the devices and manufacture it or sell it. However, it is still OK to get the plans and build one[2], which many people have done. In some ways, he was continuing the work of Nikola Tesla, in harnessing very high voltage 'static' electricity. His inventions using plasma sparks and coils to harvest ambient electricity and his demonstrations were simply amazing, with great panels of extremely bright bulbs lit with very little current in the milliamp range. In some cases it's literally a Tesla sphere, which produces resonant scalar waves, being tapped with copper plates or wires in proximity. It's not rocket science and it's so simple to replicate that many people have done so. Again, just check out YouTube.

1. Resonance Energy Methods
   http://www.free-energy-info.tuks.nl/Smith.pdf
2. Building a Smith Generator:
   https://www.aboveunity.com/thread/building-a-smith-generator/

So plasma is an extremely interesting area of science that definitely needs further investigation. It seems that using it to split water could be revolutionary in terms of just how much hydrogen output you can get with very little energy input.

Next, I'm going to give you examples of people across the world who have actually achieved water fuelled power in their vehicles, using the outlined technology. In all cases, they were discredited in one way or another and not allowed to propagate their ideas. Climate crisis, global warming, zero carbon, net zero targets, blah, blah blah. All this talk I hear on a daily basis. Every news item, every day. Meanwhile, all those people had an answer. Please start listening.

17

A few significant people

(By no means a definitive list)

"There IS no energy shortage;

only grey matter".

Donald Lee Smith

"There is energy aplenty in the wind, the sun, in flowing rivers and waterfalls... Even in a cup of water, latent there, just waiting for the adventurous... Go for it!"

Bob Boyce

There's a saying that "the proof is in the pudding" meaning that the best way of testing a theory is to see how it works in practice. On that count we have no difficulty in finding about a dozen people who, in their own unique ways, have put this technology into practice. What is certain though, is that there has been a great resistance to what they have achieved. It seems that the 'powers that be', have up to now succeeded in stamping out their innovations in favour of more profitable sources of fuel. Let's change that once and for all.

You may have noticed that my book only seems to focus on Andrija Puharić, Stan Meyer and his twin brother Stephen, who have succeeded in splitting water with an efficient form of electrolysis, but they are not the only ones and certainly not the first. It may surprise you just how many other people have too. Often they try to keep their own methods secret in order to profit. That never works as they get paid off or intimidated and quickly shut down. Even if they do succeed they are often disbelieved. Also, there are different approaches to splitting water and some are more efficient than others.

Some, like many who are experimenting with HHO, are just adding it to existing fuel systems to get a bit more mileage, but although they are producing a fair amount of oxyhydrogen gas, they are not using the most efficient technique, i.e. resonant frequency electrolysis. Some methods, like plasma electrolysis, produce massive amounts of gas, but there is little evidence of the techniques used or the results achieved. The biggest problem as I see it, is that there are no large organisations working on it that have the budgets and facilities to really make progress. However, let me introduce several people who have succeeded in one way or another. You may be surprised.

Whatever happened to stanmeyersparkplug.com?

Here's another mystery. As the name suggests, these people, after years of research, managed to replicate Stan Meyer's 'retrofit kit' along with a functioning 'water spark plug', but they seem to have disappeared completely

They produced a YouTube video claiming "We did it!" They were even offering them for sale. Now, I can't find any trace. The website is down too. Wouldn't you think that their success would now be mainstream and in use by growing numbers of people? No. They've gone. My guess is that they were paid off. However, I know quite a lot about their research, as I used the fantastic 'Wayback Machine[1]' internet archive, which takes snapshots of several previous states of a website over time so that even if they are taken down you can view those. So yes, it's still there, along with amazing detailed illustrations of their spark plug. It's actually the same as the one that Stan Meyer patented, but the important thing to know is that they got it working. If anyone knows what happened to them, please let me know.

1.   Wayback Machine - Internet Archive https://archive.org

At least a dozen people have successfully run vehicles or engines on hydrogen sourced from water on-demand. Let's see who has achieved it and how.

I am going to list a few of them here. From The USA to Australia the UK and Indonesia, experimenters and scientists have used various methods to achieve this, some preferring not to disclose details and others patenting their unique designs, some even shared openly for the sake of humanity. Most have been told to keep it to themselves or become actively intimidated. But the truth remains. It is possible and has been reproduced by many individuals over the years across the globe, not just Puharić and the Meyer twins whom I have been focussing on, but many others. For example, there's a video of a Citroën DS from way back in 1974 running on water alone, but beyond the video clip, I cannot find any more details of it. Something as sensational as that should now be common knowledge not lost in the midst of time.

Most recently, many people post videos on YouTube showing the massive production of HHO gas, a combination of hydrogen and oxygen. These often show an output some 20-30 times that of Faraday's initial results, (from nearly 200 years ago), with the same or even less input current, which is often erroneously quoted by those who would have us believe that the process cannot be efficient and therefore not worth a second look. Here follows a summary of the achievements of various people that you may like to look into yourselves. (I will be going into much more detail about the work of all of these people in book two and on the hydod.com website).

Editors note: Wikipedia has very kindly provided a list of "pseudoscientific water fuel inventions" (Let them have their opinions, I believe in free speech and I'll prove them wrong in the end. Don't forget anyone can edit this information, even government officials or those with financial interests in the fossil fuel industry).

Here are a few more people who have converted cars, motorbikes, boats and generators to run on hydrogen from water. The first, a Filipino, was put in prison and charged with fraud (that will do it, nobody will believe him then!). However, the second man presented here, being a highly respected NASA scientist, was simply ordered not to sell any of his converted cars in the state of Tennessee (annual oil output 165,000 barrels!).

Daniel Dingel

Daniel Dingel (1928 - 2010), a Filipino engineer from San Fernando, after taking thirty years to develop a water-fuelled car, was thrown into prison for twenty years for his efforts. That's a tragedy, not just for him, but for the whole of humanity in my opinion. He called his invention a 'hydrogen reactor' and most recently used it to power a 1996 Toyota Corolla. He never revealed the secret of his invention, but he did have a worldwide patent on it[1]. In that, he claimed that "*The system can be adapted, not only to automobiles, but to homes, offices, aeroplanes, jets, boats, and power utilities.*"

"The System produces hydrogen gas and oxygen from water on demand. Water is separated into hydrogen and oxygen gas using a unique combination of metallurgical, electrical and design inventions. The hydrogen gas is then burned in an internal combustion engine, what hydrogen is not burned is returned to the System with oxygen and is returned to the water storage compartment. The hydrogen provides the explosion, like octane, to propel the cylinders in the internal combustion engine".

Daniel Dingel

Daniel Dingel with his water-fuelled, hydrogen-powered 1995 Toyota Corolla

1. Water-powered fuel cell Patent: https://bit.ly/3t8HMR8

His device produces hydrogen gas and oxygen from water on demand, so no storage for hydrogen is needed, thus eliminating the risk of explosion in the event of accidental collision.

The system processes ordinary water, separating the hydrogen and oxygen in a large rectangular water fuel cell that has a honeycomb appearance, it then utilizes the hydrogen as its fuel.

Similar to many HHO installations, the gas was fed into the air intake and the vehicle was started using 30 ml of regular fuel and then switched over to 100% water fuel immediately afterwards. This is because hydrogen isn't available until the engine was running and electrolysis production is already underway, (although the latest Toyota Corolla prototype no longer needed gasoline to start the engine).

Dingle is known as a vocal critic of Filipino government officials and scientists who have refused to support his invention. The Philippines' Department of Science and Technology, in turn, has since declared his invention a hoax. However, Testing by reputable laboratories such as the Philippine Institute of Pure and Applied Chemistry (PIPAC) at the Ateneo de Manila Campus was successfully conducted. The year after, the TUV Rheinland Group through TUV Rheinland Taiwan Ltd. sent one of its engineers to analyse the system. Also, BMW was very interested in his concept except he wasn't happy to reveal its secrets to them, as he thought they might just steal the idea. He did however take his car to the USA and toured extensively to demonstrate it.

Dingel says "I don't want to think ill of anybody. I just want to make sure that these inventions get into hands who will not use them for their own selfish motives. I didn't work 14 years day and night to come up with something for rich businessmen. I was able to come up with this car because I have always wanted to make life better for people, especially the poor. I don't want to see hungry people anymore. We've suffered enough." Sadly, though, Dingle eventually accepted a "buy-out" and after he died, everything he had worked on mysteriously disappeared, including the entire contents of the website dinglefoundation.com.

For much more, see this website[1] which has a great deal of information about him and all aspects of his invention.

---

1.   https://fuel-efficient-vehicles.org/energy-news/?page_id=928

Herman P. Anderson
1918-2004

Herman believed in a Hydrogen Future and FREE fuel!

Herman Anderson was a top engineer and pilot. In World War II he served as a fighter pilot, including as a flying instructor. He was also a test pilot for the US Airforce, collaborating on top-secret projects along with NASA, including the first US satellite in space, the Lockheed SR-71 Blackbird (Mach 3 strategic reconnaissance aircraft), the Stealth Fighter and stealth Bomber and the Starwars program. He worked with Dr Wernher von Braun testing hydrogen-powered rocket engines, and he also worked with engineers at Lockheed Martin's Advanced Development Programs (ADP) (aka 'Skunk Works'), the Jet Propulsion Laboratory (JPL) and California Institute of Technology (Caltech).

So, just to make it absolutely clear, he was no pseudoscientist!

In 1971 he converted his Ford LTD V-8 (manufactured between 1965 and 1986) to run on hydrogen from water electrolysis. A car which still exists and is on display in the Water Fuel Museum in Lexington, Kentucky. He explained that ambient air was mixed with just hydrogen (not oxygen) and a micron-sized water mist (to mimic gasoline's burn), was introduced into the engine's intake, just as propane gas is on a standard propane conversion. He said he got more power than gasoline, as well as 38 miles per gallon of water. He also converted a Chevy Cavalier to run on water.

Herman with his water-powered Chevy Cavalier[1].

He was allowed to drive it, but could not sell any, or manufacture any in the State of Tennessee (Total annual crude oil production 165,000 barrels).

1. https://waterpoweredcar.com/herman.html

Cynicism, scepticism and disbelief.

When this inventor in Brazil converted his motorcycle to run solely on water, it was on a local news programme on AP Television in Sao Paulo, Brazil, now a YouTube video[1], but just listen to the cynicism with which it was presented. The female presenter says "*...with a little bit of knowledge in mechanics and **a lot of fantasy***" I bet she has no idea what a 'resonant frequency' is or that all fossil fuels, as well as water, are composed of hydrogen. Later in the clip, an 'expert' who obviously doesn't have a clue about its operation says that although he has a great deal of respect for the guy it could never work. I quote...

"Brazilian inventor (Ricardo Azevedo) builds a water-powered motorcycle"

"Moto Power H_2O" - Water-Powered Motorbike From Brazil.

With a bit of knowledge in mechanics and a lot of fantasy, a Brazilian civil servant has created, in his garage in São Paulo, what he calls the "Moto Power H_2O". He has transformed his 1993 Honda NX 200 into a kind of water-powered motorcycle. The bike uses a car battery to produce electricity, and by the process of electrolysis, separates hydrogen from water molecules Ricardo Azevedo, inventor: "This is a device that I named "Moto Power H_2O", which breaks apart the water molecules transforming them into oxygen and hydrogen. The hydrogen comes out in larger quantities and then I use this hydrogen to run the motorcycle engine. ... it makes over 500 km (300 miles) per litre of water." "We are here in the Tiete riverside, one of the most polluted rivers of Brazil and of the world. I did the experiment with distilled water, which is the most suitable, but like every researcher, I also did the experiment with the polluted water from Tietê River. And for my surprise, the polluted water from Tietê River worked as much as the ideal water. So, now I use the Tietê River water as fuel in my motorcycle."

Euronews news channel.

1. "Moto Power H2O" - Water-Powered Motorbike From Brazil 10 Feb 2016: https://www.youtube.com/watch?v=wXkx7i7kOtM

"Every sort of new device or any work has to be considered with every respect. I mean, I have all respect for these sort of people that out of concern start tinkering with the things they have at home and then came up with something that works. The only thing is that we need to inform the public that there is no magic bullet to solve the energy crisis. I mean, what is happening is an electrical vehicle, it is not running on water, it is running on electricity on the battery.

Marcelo Alves,
Mechanical Engineering Professor at the University of São Paulo

Must be running on the battery? Really? Do you think that would work? Wow! Meanwhile, the guy is happy enough to take a sip of water and put the rest in the tank and ride off. After all, he no longer needs to pay hard cash for filthy fossil fuel.

Zach West motorcycle.

An American called Zach West converted a 250 cc motorcycle to run on water. He made the system using simple, easy-to-obtain components and it follows most of the principles outlined in this book, namely a 6-volt electrolyser with eight pairs of electrodes. with a PWM input (Pulse Width Modulator (also known as a "DC Motor speed controller")). One difference is that he separated the oxygen and hydrogen so that just hydrogen is utilised after compressing it to 30 psi (pounds per square inch) and storing it in a small tank. The reason for this is to give a boost for acceleration from a standing start. A tank is necessary as compressing ionised gas within the electrolyser would be dangerous and likely to explode.

The construction is based on a pair of electrodes made from spiral-wound coils within a 2-inch (50 mm) by ten-inch (250 mm) tall, plastic pipe. There is a bank of 8 of these in the electrolyser. The coils are made from 5-inch (125 mm) 316L-grade stainless steel shim stock which is very thin so is very easy to cut and work with.

Construction details are here[1]. The motor controller can be obtained from Hydrogen Garage in America[2].

1. Zach West's Water-powered Motorcycle construction details: https://bit.ly/3wt5NDm
2. Hydrogen Garage in America: https://waterpoweredcar.com/hydrobooster2.html

Citroen DS (1970s)

There's a YouTube video[1] showing a French Citroen DS apparently running on water, but I can't find any more information about it. This is very odd considering that it was achieving this at the height of the Arab oil embargo, when Arab members of the Organization of Petroleum Exporting Countries (OPEC), caused oil prices to increase massively, during the 1973 Arab-Israeli War. This is what it says in the notes below the video.

"Can cruise up to 70 mph on water" "The engine can also power a generator, to provide light and heat, what's more, it's pollution-free" "The car has already run for well over 1000 miles and the promoters say the scheme is practical and a commercial proposition. Fuel 'on tap' in fact".

Inventor Archie H. Blue (1972) The man from Down Under

Another water car inventor whom no one knows about.

Archie Blue, from Christchurch, New Zealand, was granted a worldwide patent in 1972, for his Electrolytic cell[2]. It was featured in the book "Suppressed Inventions & Other Discoveries" by Jonathan Eisen[3]. He fitted the device to a small van and using only water as fuel, demonstrated it to the public and experts on many occasions. Michael Kemp, the motoring journalist for the UK's Daily Mail newspaper, drove the van himself and reported that it was "Lively and powerful" (although his device only used 1½ amps). Like Stan Meyer, he refused big money for the invention but was also met with a huge amount of scepticism. Sadly, when he died, his family not understanding what it was, discarded all his equipment at the local dump.

---

1. Citroen DS on Hydrogen from water. Already invented & running in the seventies. https://www.youtube.com/watch?v=79F_w9Ze_xo
2. Electrolytic Cell https://patents.google.com/patent/US4124463A/en?oq=4124463
3. "Suppressed Inventions & Other Discoveries": https://amzn.to/3N03kHO

215

Inventor Francisco Pacheco (d. 1992)

Bi-polar auto electrolytic hydrogen generator

Inventor Francisco Pacheco from Bolivia who later became a US citizen, was granted a worldwide patent[1] for his 'hydrogen generator', which is basically an electrolyser consisting of magnesium plates (a key element of the system) and aluminium plates separated by permeable membranes and seawater as an electrolyte. It claims to produce hydrogen of 99.98% purity. Important to note is that despite the seawater containing sodium chloride, no detectable chlorine gas is produced in the process. However this system does consume magnesium, but still with zero pollution and the cost of the magnesium is still far less than the cost of conventional fuel. He successfully tested the device in a car[2], a motorcycle, lawn mower and torch and in July 1974 successfully ran a 26 ft powerboat for nine hours using its inexhaustible supply of seawater!

"Since the cell generates the electric energy for the electrolysis, the cell operates as an auto-electrolysis device requiring no external energy source".

Francisco Pacheco

Due to the presence of the saltwater this device **needs no external power supply** as it works as a Voltaic, or Galvanic cell. This is why it's called a '*bi-polar auto* electrolytic hydrogen generator'. Another system based on the same invention provided electric energy and fuel for cooking and heating in his home in West Milford, New Jersey. A demonstration of this system was witnessed by the New Jersey Commissioner of Energy and staff, who did not help to take it any further. He made many other attempts to get his device accepted by mainstream authorities and the general public but was met with considerable resistance.

Francisco was committed to his belief that something had to be done about the damage that was being done to the planet and people's health by the continued use of fossil-fuels and spent about 46 years unsuccessfully trying to get his technology accepted across America, his adopted country.

1. https://patents.google.com/patent/US5089107A/en?oq=5089107
2. Gas-operated internal combustion engine:
   https://patents.google.com/patent/US3648668A/en?

Gas-operated internal combustion engine

"A gas-operated internal combustion engine for driving a vehicle, the vehicle carrying a hydrogen gas generator including a magnesium electrode immersed in a salt-water electrolyte. The rate of gas evolution in the generator is controlled as a function of demand, the gas being intermixed with air and supplied to the cylinders of the engine."

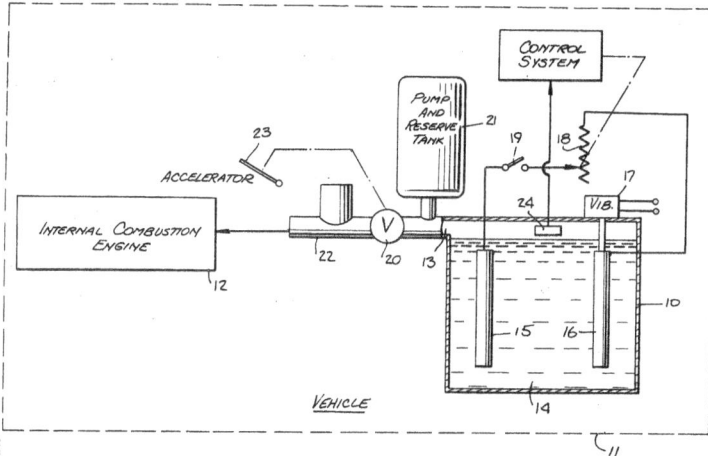

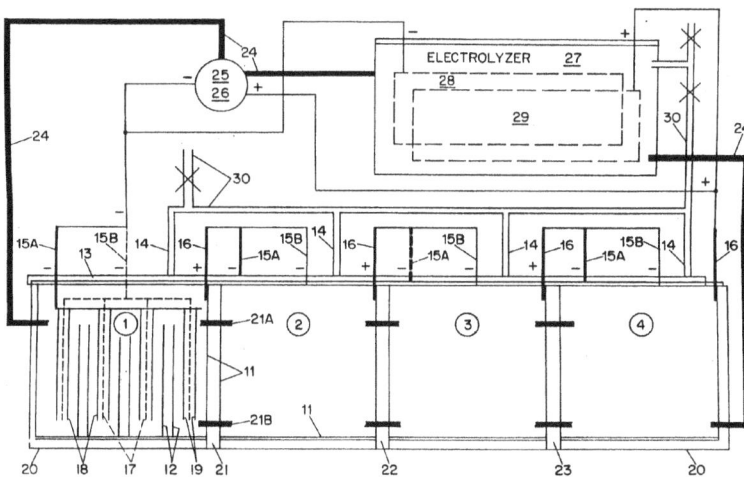

FIG. 2

The Francisco Pacheco Hydrogen Generator

"An autoelectrolytic hydrogen generator system constituted by one or a plurality of similar cells wherein a galvanic arrangement of magnesium and aluminum plates of sacrificial elements as anode; stainless steel as cathode and sea water as electrolyte, by its very nature is made to develop a voltage when connected in short circuit causing a current to flow within the system and hydrogen production of hydrogen in situ and on demand by the electrolytic action at one pole, the cathode, and additional hydrogen by the electrochemical reaction at the other pole, the anode. Surplus electric energy of the system applied to a optional electrolyzer will also be made to produce additional hydrogen at its two sacrificial aluminum electrodes".

From the patent:
https://patents.google.com/patent/US5089107A/en?oq=5089107

The genius of this invention is that he has successfully combined three principles into one device. It works like a battery with the potential produced by two different metals and the salt water catalyst, using this internally produced current to create hydrogen by electrolysis and the active metals also produce even more hydrogen. See articles here[1].

He received 2 Hall of Fame Awards in 1978 and 1979 at the Massachusetts headquarters of the Inventor's Club of America.

Francisco Pacheco's generator was successfully tested at the New Jersey Gollob Analytical Service Corporation Labs in September of 1973, also the Aesop Institute analysed the generator and wrote this report.

"I have read the literature relating to Pacheco's hydrogen generator. In my opinion, there is no reason why it ought not work as described. Basically, he has combined in one device three very simple chemical principles;

a) *The use of active metals to produce hydrogen from water,*
b) *The differing electrical potential of two metals to produce an electrical current,*
c) *The use of electrical current to produce hydrogen from water by electrolysis. All the ideas are well known; they simply haven't been put together this way before. It is so simple as to be elegant."*

Nan Waters, consulting chemist 1979

1. http://www.rexresearch.com/pacheco/pacheco.htm
    https://fuel-efficient-vehicles.org/energy-news/?page_id=926

Converting a generator to run on hydrogen from water

When you think about it, there are already engines and generators which are effectively standing engines, already set up to run on gas, albeit LPG gas (Liquid Petroleum Gas). So little adjustment is required to convert them to hydrogen power. However, hydrogen is not the same as LPG and the differences matter. But, having said that, unlike an engine in a vehicle, you do not have to worry about variations in the rev range to adjust to road speed and acceleration! There's a comprehensive guide freely shared by David Quirey[1]

'Hidden Technology' - Running a generator on water.

This excellent half-hour video "Hidden Technology" shows a portable generator running on just water from an electrolyser[1]. They also show very clear steps of the entire construction.

It consists of a set of twenty stainless steel plates, suspended inside a water filter housing held by two long, threaded, bolts and arranged so that every alternate one is positive or negative. It's a very elegant, simple, but effective design. A second filter housing is used as a 'bubbler', which helps to prevent flashbacks, as hydrogen can't pass back through the water. They go on to mount it onto the side of a 3000-watt portable generator and demonstrate it working inside and remotely. (And still, there are disbelievers in the comments). Along with possible other adjustments, they cover up half of the air intake to balance combustion as HHO is part oxygen.

The added 'HH+' electrolyte being pink, suggests that it is based on cobalt chloride (polyoxometalate, precipitated as a barium salt). That with a binding of carbon paste can create a cheap but highly effective catalyst. However, the actual components of the mixture used here are kept secret. I suspect that what will happen, before long, is that they will be offered a substantial amount of money to stop what they are doing, subsequently take the video down and disappear for good. Either that, or they will try to patent it and the patent will be bought out and hidden for ever[2].

1. Running a Generator on Water: https://bit.ly/3wUisQj
2. THE GREATEST INVENTION: WATER AS FUEL!
    The secret of the HH+ compound to boost hydrolysis - YouTube: https://bit.ly/3wx9sjK

Major contributions

Here are a few people who, in their own way, are making huge contributions to this technology.

There are many others, but this book is already getting on for three hundred pages. With more and more urgency I need to get this book finished and published. I heard in the news yesterday, that here in the UK, due to rapidly rising fuel costs causing 'fuel poverty', despite oil and utility companies making record profits, cases of food poisoning were rising in children because their parents were turning off refrigerators and freezers overnight to save money. The answer to the problem has been under our noses all along. It's just that the big players couldn't profit from it.

Bob Boyce

Bob had an electronics business down in south Florida where he owned and sponsored a small boat-race team through his business, starting in 1988. He had a machine shop behind his business, where he did engine work. He worked on engines for other racers and a local minisub research outfit which was building surface-running drone-type boats for the DEA. He delved into hydrogen research and started building small electrolysers using distilled water mixed with an electrolyte. He then resonated the plates to improve the efficiency of the units. He discovered that with the right frequencies, He was able to generate 'monatomic' Hydrogen and Oxygen rather than the more common 'diatomic' versions of these gases. When the 'monatomic' gases are burnt, they produce about four times the energy output produced by burning the more common diatomic version of these gases.

About 4% of diatomic Hydrogen in air is needed to produce the same power as petrol, while slightly less than 1% of monatomic Hydrogen in air is needed for the same power. The only drawback is that when stored at pressure, monatomic hydrogen reverts to its more common diatomic form. To avoid this, the gas must be produced on-demand and used right away. Bob used modified Liquid Petroleum carburettors on the boat engines to let them run directly on the gas produced by his electrolysers. Bob also converted an old Chrysler car with a slant six-cylinder engine to run on the hydrogen set-up and tested it in his workshop. He replaced the factory ignition with a high-energy dual coil system and added an optical pickup to the crankshaft at the oil pump drive tang to allow external ignition timing adjustment. He used Bosch Platinum series spark plugs.

Bob never published anything about what he was working on, and he always stated that his boats were running on hydrogen fuel, which was allowed. Many years later that he found that what he had stumbled on was already discovered and known as "Browns Gas" (see chapter 9), and there were companies selling the equipment and plans to make it.

Bob Boyce is a true humanitarian and has placed his system in the public domain in the hopes of helping the planet reverse global warming, end oil wars, and deliver free energy to the poor peoples of the world.

"Water IS life… and so much more… So, why all of the excitement about common H_2O? Water is the glue that binds all of life together. Without it, we would not exist. With it, not only do we survive, but we may have a solution to our growing global energy problem".

Bob Boyce

100 cell superefficient electrolyser.

Bob Boyce's superefficient electrolyser is constructed of 100 cells (101 plates). If you want full construction details, then please take a look at Patrick Kelly's D9.pdf [1] (or Pages 6 - 49 of his eBook 'Practical Guide to Free Energy Devices[2]'). It has full details, diagrams as well as the electronic circuits required. Patrick Kelly's superb free 3024-page guide is open-source and he shares the same mission to get this technology out into the public domain, as does Bob Boyce. (The D9.pdf is just a small part of the same publication).

"Water can be transformed into a perfect energy supply. It is abundant, non-polluting, and eternal in nature. You split it efficiently and combust it efficiently. After

1. https://waterpoweredcar.com/pdf.files/D9.pdf
2. http://vrr.dyndns.biz/Docs/OLE/FreeEnergy/PJKbook.pdf

harvesting that released energy, you again have H_2O as the by-product. Hard to beat!

Will the greed of big oil and big business ever be satisfied enough that they can stop destroying this planet? I seriously doubt it.

There is energy aplenty in the wind, the sun, in flowing rivers and waterfalls… Even in a cup of water, latent there, just waiting for the adventurous… Go for it!

Bob Boyce

Dave Lawton

Phase lock loop circuit (along with bifilar coil)

One person worth mentioning here is Dave Lawton, from Wales in the UK. His pulse-generating circuits for water-splitting are getting a lot of interest online. He has designed an 'auto-tune' PWM circuit around a pair of common 555 timer/oscillator chips. They have been shown to be very successful in achieving a large output of gas with very low current. (e.g. 230 volts with only 1.5 Amps maximum). The importance of this is that one reason why people have difficulty getting water-splitting to work successfully is that there are too many variables. Pressure, temperature, water composition and quality, fuel cell size, plate type and gap size, etc., and along with all that, trying to find the right voltage, current and waveform to bear down on the invisible microscopic water molecules is a bit like trying to find a needle on Everest. Therefore, if you have at least an 'auto-tune' circuit, it almost sounds like an unfair advantage. Apparently, he was successful in demonstrating the operation of a portable generator running solely on water. Although I am also in the UK, I have unfortunately been unsuccessful in contacting him directly. The last post he made on Facebook was two years ago, which was the beginning of the Covid-19 pandemic. I don't want to make any assumptions there and I hope that has nothing to do with his silence. So, Dave, if you read this please get in touch, or if anyone else knows anything please let me know.

Ravi Raju

Ravi Raju from India produced the results of his research paper on 'pulsed DC resonant systems' with water fuel cells, in an excellent 41-page guide called simply 'ravi.pdf'[1] (detailing how to construct a fuel cell containing a set of tubes much like those used in Stanley Meyer's patent application). In it, he shows some very interesting results, not least that no current was required as the voltage potential alone did the work of producing hydrogen. After disclosing his gas flow test on YouTube, he was unlawfully intimidated and threatened which was intended to interfere with his research work. The same old story sadly.

He concluded:

"Nothing about this process involved in the Water Fuel Cell resembles electrolysis. There is no electrolyte used; there is NO current admitted to the water, in a proper system. No heating occurs in the water as it produces gas. The gas produced is Hydroxy (aka HHO or "Brown's Gas"), not differentiated Hydrogen & Oxygen; and all of the work is performed by voltage potential alone".

Petkov (Valentin Petkov /valyonpz/) electronics engineer - YouTuber

I'd like to mention 'Petkov' here. He's a YouTuber[2] from Romania, with a considerable subscriber base of 3.48K members[2]. As a member of open-source-energy.org, he has contributed a great deal to the research into the use of water as a fuel source. He has posted many videos of his research, openly sharing his analytical conclusions. With shots of his oscilloscope, you can see that he has accurately replicated the waveforms of Puharić, Stan and Stephen Meyer, along with explanations and details of the equipment and components that he needed to achieve those results. The last video he posted was just one month ago 'Stephen Meyer`s 3-Electrode Water Fuel Cell[3]' (as covered in chapter 13).

1. Research Paper on Ravi's Water Fuel cell Replication:
   http://www.tuks.nl/WFCProject/pdf/ravi.pdf
2. https://www.youtube.com/c/valyonpz
3. https://www.youtube.com/watch?v=vErg4LVz_b8

Alex Petty

The excellent website, alexpetty.com, 'Energy Research Journal' by Alex Petty, includes very clear and concise explanations of Puharić's water-splitting technique as well as Meyer's, and even includes the mappings between Meyer's hydrogen GMS Cards and patent schematics. Working in his basement, he has replicated Stanley Meyer's hardware and openly shares his results.

He also runs a YouTube channel[1], posting videos of Puharić/Meyer waveforms and other proof-of-concept research.

Daniel Donatelli Secure Supplies

Daniel Donatelli[4], another member of Open-source-energy.org forum, is actively researching Puharić and Meyer's water-splitting technology and goes as far as freely sharing circuits as well as offering fully functioning equipment for sale. He also posts 3D design files ('STL') on Thingiverse[2], which you can freely download and print (or have printed), in order to experiment yourself. Incredibly, that also includes Stan Meyer's water fuel injector parts! He has also posted circuit board designs on PCBWay[3] of Meyer's digital control boards.

He posts frequently on his YouTube channel[5] too.

FreeFromFuel.com

www.FreeFromFuel.com

This is a great little workbook from 'Living Oneness Vast Evolution' that I would like to recommend. (And no, I'm not on commission here). It's a neat little DIY manual full of design templates and loads of useful information, plus a DVD with many hours of instructional videos. The website has downloadable files covering making dry cells, flame arrestors and a hydrogen welding torch,

1. https://www.youtube.com/c/AlexPetty999
2. https://www.thingiverse.com/search?q=Daniel+Donatelli
3. PCBWay https://www.pcbway.com/project member/?bmbno=E9AF6EAA-8A83-4F
4. https://danieldonatelli.wixsite.com/hydrogen-power-gas/
5. https://www.youtube.com/c/SecureSuppliesLimited

as well as hydrogen sand heaters for domestic heating. These people seem quite experienced at making all things to do with hydroxy gas production and so are worth listening to. I can also recommend their YouTube videos (perhaps as a taster).

One caveat though. Their projects don't seem to be using a resonant frequency to optimise production. Nevertheless, they do give a nod to Stanley Meyer and mention that using a resonant frequency should increase output.

Patrick J. Kelly

Patrick Kelly has produced this superb, free, **3,024-page** eBook ('pdf' format), (For comparison, Leo Tolstoy's 'War and Peace' is only 1,225 pages!).

It's called the 'Practical Guide to Free Energy Devices[2] and has full details, diagrams and electronic circuits, including many water-fuel devices and is also open-source. He shares the same mission to get this technology out into the public domain. However, it has more than 400 bookmarks to speed up your access to any section of interest. (It has also been split into separate chapters for convenience). He also produces YouTube videos (see his channel here: https://www.youtube.com/user/TheEngpjk)

http://vrr.dyndns.biz/Docs/OLE/FreeEnergy/PJKbook.pdf

Russ Gries

Russ Gries[1] runs a very lively forum[2] with like-minded individuals working on hydrogen-on-demand projects to crack the code to releasing hydrogen from water and associated technologies. He also runs a highly informative YouTube channel[3] about all things to do with using water as a fuel and is obviously keen to share all his research along with Alex Petty, who also has a wonderfully informative site[4] of his own.

---

1. Russ Gries www.rwgresearch.com
2. www.open-source-energy.org
3. https://www.youtube.com/user/rwg42985
4. Alex Petty www.alexpetty.com

My aim and object is to make this amazing and life-giving technology as public and as available as I can. Many others have been working on drawing hydrogen from water as a fuel for many years, but it seems as if their research has either been dismissed or quenched. My purpose is to make it so public that no one can ignore it any longer, and we can change this polluted world forever.

In recognition of all these amazing people (including the aforementioned Puharić and the Meyer twins), let's see what we can find out about their work and learn by it. I have and I can tell you that I am totally convinced that they are all genuine. Let's take the opportunity to bring this amazing technology out into the world. I have taken the trouble of gathering links to many external sources, especially their patents in Appendix 2.

Epilogue

Filthy Oil

or

Clean Water?

My mission is to bring this technology out into the world in a way in which it will be impossible to hide it any longer. Unfortunately, it's not as if you can go and buy a water fuelled hydrogen-powered car from a high street dealer. Also, it seems that many people have got most of the parts but nobody has succeeded in bringing it to market. I believe that I've explained why that has been the case up to now. I hope that I have convinced you that the technology is valid and worth pursuing. What is needed is for large organisations with the resources and personnel sufficient to take it further. Let us hope that we are not too late.

Why Renewable Energy sources are not the answer

Fuel cells and hybrids are not the answer

There is a huge amount of money being poured into fuel cell and hybrid solutions by top companies around the world, especially vehicle manufacturers, but in my view, they are all backing the wrong horse. Hydrogen on demand from water supersedes the current fuel cell technology for several key reasons.

A fuel cell is an electrochemical device that converts hydrogen and oxygen to electricity to power electric motors which propel the vehicle. However, there are major issues with this technology, mainly that it requires a specialised and substantial holding tank for the hydrogen which needs replenishing at a refuelling station and the use of large batteries to store the electricity which is produced. The batteries are usually expensive lithium versions too.

Hydrogen storage is a huge issue as it can't be easily liquefied like LPG and it is such a fine element that it can pass through steel. There are places in the world such as California, where around 40 hydrogen refuelling stations have been built, but they are still relatively thin on the ground.

Stephen Meyer (see chapter 13), twin brother of Stanley Meyer, the well-known water-powered car designer (see chapter 11), most probably spooked by his brother's death, has since channelled his energies into creating hydrogen filling stations, based on exactly the same principles. He has patented a device using the same resonant frequency technology as his twin brother Stanley Meyer to release hydrogen from water. In other words, it uses the exact same hydrogen on-demand technology to improve efficiency. (He is also using the technology to purify water on a large scale).

The likes of Ford, Mercedes, BMW, Honda, Toyota, Hyundai, etc. are all putting in a great deal of research and development in this emerging hydrogen technology. (Although it's not a new technology, being invented in 1839 by William Grove). It's a technology being increasingly used as a solution, not just for cars, but also for buses, fork-lift trucks and generators and is even being considered for aircraft, large and small. It is indeed a zero-carbon solution, with no emissions except water. They may solve the range limit issues of electric vehicles, (leading to 'range anxiety') being capable of 300 miles, or more, between hydrogen refills, but still need a special hydrogen filling station somewhere to replenish.

Elon Musk, the electric car manufacturer, calls hydrogen fuel cells 'mind-bogglingly stupid', but then so are electric cars in my opinion, as they are expensive, have a limited range with charging issues, and *still mostly use electricity sourced from environmentally unfriendly fossil fuels*, which completely defeats the objective. (I'm sure there are people who think they are 'saving the planet' when, in reality, they are charging their vehicle with electricity from gas- or coal-fired power stations.) Hybrids are the worst of both worlds, except when one fails, at least you have the other. They still use fossil fuel engines and expensive battery/motor combinations.

Hydrogen on demand doesn't require onboard hydrogen storage tanks or specialist refuelling stations, as the hydrogen is produced to meet the demands of the engine only when required. A fuel cell vehicle is essentially an electric vehicle whereas existing ICE (Internal Combustion Engine) technology can be adapted to run on hydrogen sourced from water alone. Indeed existing vehicles can, and have been, adapted to run on hydrogen, similar to LPG gas. After all, fossil fuels are turned into gas vapour in the cylinders of an engine (perhaps why Americans call petrol 'gas'), so it's not much different. Modern engines use computer-controlled electronic ignition too, so the technology for hydrogen on demand is not radically different.

It's time that these large organisations started looking at hydrogen on-demand as a far better solution, before billions more money is wasted in stranded assets and redundant, obsolete technology.

Nuclear energy is not the answer

Although it's often misleadingly portrayed as a 'clean' 'economical' energy source, nuclear energy is actually a by-product of the cooling system of nuclear fusion, mainly used to produce plutonium for nuclear weapon systems How something dangerously radioactive can be described as 'clean', I'll never know, not to forget the terrible accidents that inevitably occur. How can this be a solution to help the environment considering what happened at 3-mile island in Pennsylvania, USA, Chernobyl in the Ukraine, Fukushima in Japan, and SL-1 Nuclear Reactor in Idaho, United States?

As I have mentioned elsewhere, the truth is that electricity is only a by-product of nuclear power from the cooling water, and the main purpose of nuclear power stations is to produce plutonium for nuclear warheads, (far from 'clean' and far from 'cheap'!). Just for the record, Iran was accused of building nuclear 'power stations' for that very purpose!

Nuclear reactors also require massive amounts of cooling water. Those inland waste millions of gallons of valuable fresh water and damage ecosystems in the process, possibly also with radioactive leaks.

Net-zero offsetting isn't the answer

'Net-zero offsetting' is effectively saying that some polluting is OK as long as carbon is removed in other ways. However, the pollution that remains is still a health hazard and still adversely affects the environment.

Solar power isn't the answer

Solar is expensive to install but has its place. However, the sun only shines in the daytime and during fine weather.

There's a new trend of covering farm fields with hundreds of solar panels, which is not good for birds or other wildlife and is something of an eyesore.

Land is too precious to completely cover in this way and it really is required for food production, whether that be for crops or farm animals. As the population grows, land is growing scarce.

Wind power is not the answer

Wind power also has major issues. Again, the weather is fickle and sporadic. The giant turbines are difficult to manufacture, very awkward to transport to their rural locations and are a blot on the landscape. They are also very expensive to maintain, especially those out at sea. They have been known to lose their giant blades, which is of course, very dangerous, not only that, but in high winds, they can overheat and catch fire. (A 2020 article in the Wind Power Engineering Magazine estimated that as many as 1 in 2,000 wind turbines catch fire each year, resulting in significant structural damage or total loss).

Solar, wind and wave power are not much use in moving vehicles either, (with the exception of solar, which can be useful to charge batteries under certain circumstances, but insufficient for locomotive power). If we are going to use renewables to make hydrogen, then we have the problem of storing it.

The problems of Storing hydrogen.

It's possible of course to split water molecules to release hydrogen using carbon zero technology such as wind and wave or solar power. This leaves you with the problem of storing the hydrogen which can be expensive and possibly dangerous. Hydrogen can be safely stored, particularly for vehicle use by storing within a metal 'hydride' lattice. These particular metals, however, can be rare and hard to come by. Lithium hydride (Lithium 6) is a metal which can be used for hydrogen storage, but it's also required for making thermonuclear bombs, like the hydrogen bomb. For that reason, it's actually illegal to buy or sell it in the USA even if it's not for such purposes.

Hydrogen-on-demand from water is a solution to all these problems.
- Water is freely available everywhere
- No dangerous (and heavy) hydrogen storage is necessary
- Nonpolluting, with plain water the only emissions
- Inexpensive technology (can even be retrofitted to existing engines)
- Requires little, if any, infrastructure

Geography and Politics

Oil for arms

One major issue is that often we are dependent on politically unstable and unfriendly regions of the globe for oil and gas supplies. The revenue from these supplies directly funds their war aims. In fact, many wars are directly related to the vast fossil fuel reserves that we all hitherto have depended on. By using water as a source of fuel, all these problems may disappear overnight.

It's time to stop depending on oil and gas from rogue nations.

It would surely be a massive benefit to governments and their people to switch to water instead of oil and gas. Just look at the global issues that we have currently. Spiralling global costs of oil and gas leading to 'fuel poverty'. Reliance on Russian gas, France threatening the UK that it's going to cut the power off to the Channel Islands over fishing rights. Piracy of oil tankers off Somalia. Incidents with shipping off Iran. Constant problems with the Middle East and wars, like those of Iraq, Afghanistan and Libya.

Oxygen from plain water

In this book I focus a lot on using a combination of hydrogen and oxygen sourced economically from water but let's not forget how valuable it could be to be able to produce oxygen on-demand for other purposes such as medical needs. Usually, oxygen for medical centres and hospitals is delivered, at great expense, as a liquid under pressure in special tanks (which is also a hazard).

In emergency situations, such as global disaster relief, a shortage of oxygen can be a real problem.

Recently, for example, during the Covid-19 pandemic, some countries were experiencing shortages of life-saving oxygen. That is crazy considering that it can be obtained on-demand from plain water.

About me and why this has become my mission.

I am a researcher and technical author as well as a teacher/trainer. I have spent my career in technical publications which has given me an insider view of many areas of industry, particularly the utility companies. I've worked on contracts for global oil and coal companies, electricity generating companies, the water board and many other relevant organisations. Being exposed to the inner workings of these industries is often an eye-opener.

My father was a sign writer and artist but during World War II was trained with REME, the Royal Electrical Mechanical Engineers. Passing on his analytical approach to life and fine engineering knowledge, as well as a solid grounding in art and design, led me to ultimately spend my career in technical publications, which require very accurate illustrations in 3D and a very analytical mind. In primary school I was more interested in the ideas of Einstein than collecting stamps or playing football. I often played chess with the headmaster and started making complex mathematical models.

At the age of eight, I won a school prize for a project I researched on the invention of the hovercraft[1] and realised that the process of researching and writing about the subject served to consolidate my thoughts and fill in the gaps in my existing knowledge. In the future I may well do the same for other technical subjects. Perhaps this was a sign that I would go on to spend a career in technical publishing! As a boy, I loved to take things apart to discover how they worked, particularly batteries, dynamos and anything electrical.

One of my earliest jobs as a 16-year-old was a short commission to draw archaeological finds for publication at Manchester University, which ultimately led to a career in technical graphics and illustration and on to technical writing, mainly for power stations, but also for coal mining equipment as well as nuclear physics and electronic equipment.

One thing very relevant to this project is that my home in Sale, Manchester was very close to the Shell petrochemical works at Partington, (a place where I even had a stint working a temporary manual job). I also had a fascination with water mills (early alternative, water-powered technology!) resulting ultimately in me making home in one in Devon in recent years.

I was introduced to computing at Bradford university in the early '80s when computers were the size of a whole department and programs were stored on punch paper tape. Later in life, after working as an assistant publisher at the University of Bath, I changed my career direction and became an instructor of computer software and ultimately became an Adobe certified instructor in InDesign, the industry standard for publishing.

I subsequently studied for a science degree with the Open University. Research is one skill that has followed through, possibly due to a great love of libraries, lateral thinking skills and massively helped by modern internet resources.

One thing that I have always wanted since an early age is a VW beach buggy, so Stanley Meyer's water-fueled hydrogen-powered car really caught my attention. (see chapter 11, p140). I have had half a dozen air-cooled cars over the years, mainly Citroëns though, not VW, but I did enjoy the ownership of one VW campervan, which had a 2 litre Porsche, (horizontal, flat-four cylinder), air-cooled engine. I have learned to work on many aspects of these incredibly straightforward engines over the years, gaining lots of mechanical skills and a good working knowledge of their operation. One or two of the Citroëns also had the added complication of the hydraulic suspension to deal with. I have also been involved with the operation of engines, their electrical systems and electric motors, in the course of my career as a technical author.

All of this first-hand knowledge has helped me understand the principles of operation of the internal combustion engine and associated technologies, which set me up perfectly for my research into the use of water as a source of energy. This technology isn't just for air-cooled engines, of course, but as a research platform, they are incredibly simple in operation and very accessible. Removing a VW air-cooled engine, for example, can be done in minutes, as there are only four bolts to remove it from the vehicle. They are extremely lightweight and can easily be set up and run on a workbench. All the principles of water-fuel technology can of course be applied to many other types of engines, whether air-cooled or liquid-cooled. The important consideration is that adjustments will need to be made to optimise for running on a much

1. Note: One vitally important point I found out about the hovercraft invention, during that research, is that the military, seeing its potential on the battlefield, effectively took the patent and made it top secret. So it was kept under wraps and prevented from being released into the public domain for 15 years! That was after the military realised that it had limited potential due to problems such as a lack of precise steering.

more efficient and instantly exploding hydrogen flame, coupled with a good supply of oxygen, which also massively aids combustion. One thing that won't be happening though, is the contamination and deterioration through partially burnt carbon. This is really worth taking seriously.

As a child, I was also very interested in electronics, magnetism, and anything mechanical. The invisible and yet powerful electric and magnetic fields seemed like magic to me. I was often messing around with batteries, magnets and electric motors, and pulling apart anything I could get my hands on. I needed to get inside to see how it worked or what it was made of. I'm sure this is what led me to end up choosing technical writing as a career. Trying to make sense of all this led me to spend a lot of time in public libraries and my passion for research. I have a huge preference for technical books, rarely taking the time to read novels. I always had this feeling that I was on the verge of some discovery - that all this was leading somewhere. I now believe that it has.

The devastation of crude oil pollution

During my childhood, I had first-hand experience of the devastation of crude oil pollution. Escaping from the city of Manchester, we had family holidays in the generally unspoiled beauty of the North Wales coast, particularly the island of Anglesey, situated on the edge of the Irish sea and accessed by a road bridge across the Menai Straits. It is a wild and beautiful place, but one particular beach had been devastated by an oil spill. The port of Liverpool is not far away and crude oil would be regularly delivered to the petrochemical refineries of Stanlow, in Ellesmere Port. (On the Mersey at the Tranmere Oil Terminal.) The small usually unspoiled, picturesque cove was our planned destination for a day at the beach, but to our horror there were dead seabirds everywhere, a stench of death and patches of crude oil all over the beach. We left immediately and spent hours trying to get the filthy stuff off our shoes and clothes. It was a horrifying experience.

I have also seen first-hand the devastation to our coastline and wildlife of one of the worst oil spills in history. The 'Torrey Canyon' supertanker[1] ran aground on rocks off the South West coast of the UK in 1967, spilling an estimated 25–36 million gallons (94–164 million litres) of crude oil. Hundreds of miles of British, French, Spanish and Channel Island coastlines were affected. It remains one of the worst oil spills in history. Sadly it is one of many that have happened globally, from the Exxon Valdez in Alaska in 1989 to the Deepwater

Horizon drilling platform catastrophe in the Gulf of Mexico in 2010, which also killed 11 workers and injured many others when an explosion occurred.

Growing up with smog - filthy city pollution

Pollution was not talked about much in the 1950's and 1960's. In those days all the buildings where I went to school in Manchester were black with the build-up of years of pollution. We experienced dense smog every winter. That is fog, polluted with soot from coal fire smoke and vehicle emissions. It was so bad that sometimes you could not see the opposite side of the same street. I remember riding my first cycle, a Christmas present, up and down inside the house as it was too bad to go out, with zero visibility and choking air. There were no controls on vehicle emissions at that time, no catalytic converters and no MOT (Ministry of Transport) tests to check exhaust output.

Soot, smoke and carbon monoxide from coal fires

The government introduced 'smokeless zones' to try to alleviate the problem. Up to then, we ourselves had coal deliveries to use in an open fireplace. One big advantage of coal is that it does burn slowly for many hours. The problem is, of course, that it is filthy and fills the air with sooty smoke and unseen carbon monoxide and carbon dioxide. Just handling the coal is enough to get soot everywhere. We also had a paraffin heater, which was also smelly and gave off noxious fumes. For all those reasons, we soon converted to electric fires. However, the unseen source of the electricity is a coal-fired power station! Later on, in my career, I would play a major part in their operation as a technical author, working on maintenance manuals and also an asbestos removal survey (but that is another story). After the electric bar fires, we converted again to natural gas, a far cheaper option, but still a fossil fuel with its associated emissions.

The Arab oil price hike

Another significant event in my upbringing was the aforementioned 1970s oil crisis when Middle Eastern Arab oil producers imposed an embargo around

1. Torrey Canyon oil spill 1967 https://en.wikipedia.org/wiki/Torrey_Canyon_oil_spill

1973. In the UK, this was made worse by the introduction of a 'luxury' tax or 'Value-added tax' (VAT) of 24%. This hit at about the same time as I was learning to drive. I remember being horrified at the huge chunk of my pay packet I had to hand over to fill the tank of my first car. Perhaps, for this reason, I was already looking at alternatives such as LPG (Liquid Petroleum Gas) conversions.

Alternative energy

During WWII vehicles were converted to run on LPG supplied from a huge gas bag on their roof. I have always been interested in alternative energy, such as wind power and solar panels. There was, and still is a 'Centre for Alternative Energy' in Machynlleth, North Wales, not far from where I was brought up. I was always interested in water mills too. There was one in particular, near to where I was born in Cheshire. That is the Dunham Massey sawmill. Dunham Massey was very much a part of my childhood. I knew the estate well as I used to go on scout camps there and family walks in the grounds of the manor house. I even knew the family that owned it before it was passed on to the National Trust. For those reasons I did a detailed study of the sawmill, producing a large cutaway painting of the inner workings, with its line shafts and leather drive belts. I've subsequently had a lifelong interest in watermills and even lived in one in Devon for a couple of years. Water power is clean, carbon-free, environmentally friendly energy. In the 1950's and 1960's when I was growing up, alternative energy, such as wind and solar power etc. was very much considered a fringe solution to providing power.

A fascination with steam engines

I have had a life-long interest in steam engines and traction engines and often attended the great country rallies that showcased them. It does, of course, trouble me that they produce great clouds of sooty smoke though. Again, as a child of the 1950's, I saw the tail end of their use as the main locomotive power of the railways before they switched to using diesel engines and electrified many of the lines.

Geology is another great interest of mine, particularly fossils. I learned at school how layers of ancient forests became decomposed into layers of coal, oil, and gas in the rocks beneath our countryside. As I mentioned earlier, I have also worked, in my capacity as a technical author, in several main coal-fired power stations, which are, of course, powered by steam turbines.

What about my being a petrol-head?

I am a self-confessed 'petrol-head', a lover of the great internal combustion engine, in cars, boats and even planes, but I won't have to give up my passion, as clean-burning hydrogen on-demand can work even better in our finely engineered engines. Fancy being able to run them on plain water, so not being restricted by the high price of fuel. Fancy not getting your pride and joy clogged up with filthy, sticky, dangerous carbon deposits. Fancy not risking your health with airborne microscopic cancerous particulates.

The high cost of fuel

During my life, however, I have struggled with the cost of fuel, from transport costs, including my own vehicles as well as bus, train and air fares, to household bills for gas and electricity. It's taken a major chunk out of my wages. That and the tax that's loaded on it as well.

If you think about it, as well as our own transport, all our goods are transported by road, rail, ship and air, so the cost of living is directly affected by the cost of fuel. It is my personal view that it would boost the economy if fuel was virtually free. Why? Because if everyone is not spending their hard-earned cash on fuel, they will have more disposable income to spend elsewhere. Secondly, expensive fuel will discourage people from travelling, or at least spending when they get to their destination. How does a holiday destination benefit from the huge cost that the individual has incurred to get there? The cost of transportation can make commodities very expensive. It would boost trade if we could remove that component of the price structure. Often the cost of transportation is actually prohibitive. Especially where goods are bulky or heavy.

The high cost of transporting fuel

The cost of transporting fuel itself is very high. (Not to mention the environmental and health implications). From ocean-going oil tankers to rail and road-going tankers. I've heard it said recently that about half of all shipping is transporting fuel of one kind or another. Think also about local deliveries of petrol to forecourts and heating fuel oils to homes, schools and workplaces, plus LPG gas cylinders to many who are off-grid. Coal from mines has to be transported by road and rail to power stations. Think also about the cost of many miles of gas pipelines which would not be required if hydrogen can be produced on-demand at source from freely available water. Particularly gas pipelines across fragile geopolitical landscapes which have very recently caused so much consternation. (The Russian pipelines to the EU in particular and places with local land issues like those recently in Canada where there were protests about a pipeline crossing tribal lands).

NetZero goals

Government vehicles use fuel and MPs need to travel, often globally. Our own UK Prime minister Boris Johnson famously hired a big (polluting) jet, at taxpayers' expense, just to return 400 miles back to London from the Cop26 climate conference in Glasgow, completely ignoring the whole point of it. The UK seeks to drill even more oil and gas from the North Sea and open new coal mines, undermining its position as leader of the summit. All parties face targets and NetZero goals they have jointly agreed, to reduce emissions by 50% by 2030. The biggest gain for governments will be the ease with which they will be able to meet these carbon-zero targets, if they are able to source zero-carbon hydrogen from zero-carbon water!

Bright future possibilities

Sourcing hydrogen on-demand from water would not only be a solution to the current extreme environmental crisis, it could actually be a vast improvement in our lifestyles and long overdue. Imagine a future where this simple technology of hydrogen sourced from water is taken on board. Imagine a world where pollution from the use of hydrocarbons is a thing of the past. Imagine a world where everything from transport to power generation to heating, ventilation and air conditioning is provided by hydrogen directly from plain water!

Imagine also that there is no longer a need for squabbles with oil-producing nations. Imagine an end to human causes of global climate change! This technology is proven but hidden in plain sight by greedy oil corporations and governments who want to tax filthy oil instead. There are huge benefits to our health, our climate, and our pockets. The sky would be clear (as it was in many major cities due to the recent lockdowns imposed during the Covid-19 pandemic). Pollution of our environment would cease, including from ships, trains, and planes. Environmental catastrophes would be averted such as those oil spills and fires. Any water can be used, seawater, river water, even snow! Freshwater would no longer be wasted in massive quantities by power generation plants. There would also no longer be any need for the wasteful power lines blighting the landscape (or those massive, ugly wind turbines), as we won't need massive centralised power stations to provide electricity, since we can easily generate our own at home, in our business premises or factories, or wherever it's needed.

We already have generators powered by LPG gas, including small portable ones that can be used anywhere, especially out in the field. A switch to hydrogen, sourced from water is an easy conversion and people are already achieving it. (see page 219).

Hydrogen on-demand from water will help governments meet their NetZero goals!

Zero carbon = zero pollution!

Not only is diesel or petrol prohibitively expensive, but also a massive pollutant for the air and environment - not only in use, but also in the process of transporting it to where it's needed. (Think of those massive catastrophic oil spills). Imagine a future where there is simply no pollution from transport of any kind because hydrogen sourced from water on demand contains no carbon and the by-product is simply water. Motorcycles, cars, boats, superliners, trains, farm machinery, mining equipment, construction equipment, and even aircraft. In fact, ESPECIALLY aircraft. (Some people that I speak to with that prospect, question the weight of the water, but using hydrogen you could do far more miles than you could with the equivalent aviation fuel). And transporting water is SAFE, unlike transporting hydrocarbon fuels. Remember, I am not proposing STORING or transporting hydrogen. It's produced on the fly, on-demand, when and where it's needed.

The economy.

Think also about the expense. Not just the huge environmental and health costs, but the massive cost to the consumer of oil-based products, coal, oil and gas, for heating and air conditioning, cooking, and transport. The production of food would benefit massively too, requiring freshwater, and climate-controlled greenhouses, not to mention the cost of running farm machinery day in and day out. Maybe we can all use the savings made by not having to pay for expensive fossil fuels to improve the way we manage the environment for a better cleaner world?

There is no doubt that a switch to using water for fuel will change the whole world economy. Governments may worry about a 'loss' of revenue from their tax on fossil fuels. (Although you can't 'lose' something you never had, which is a common myth). However, it doesn't mean that they will lose out for several reasons. Firstly, I believe that there would be a massive boost to the economy. People will then not have to spend such a huge amount on utility bills and travel and also on cheaper goods, so they will suddenly find a huge increase in disposable income. In short, they will spend their money on other taxable goods and services instead. Government departments will also save a huge amount on delivering services, such as waste collection and disposal and water purification and desalination. Not to mention recycling.

Reduction in the cost of travel

There may be other surprising benefits too. For example, as shipping will be so much cheaper to run (nonpolluting and safer too), that new markets could open up. Overseas travel will be cheaper. History tells how important that is for civilisation, from Ancient Egypt to the Roman Empire and trading hub cities such as Venice, not to forget the British Empire. This is one reason that China does so well in modern times, shipping goods to the rest of the world. Planes can travel lighter as there is three times as much energy in water as in fossil fuel, plus, it eliminates the risk of fire and explosion if plain water is carried instead of highly volatile fuel!

Let's hope for the sake of humanity that governments will take their blinkers off and allow this technology to flourish rather than suppressing it as they have been up to now.

We'll still need oil products.

Oil companies will still have to provide a certain amount of oil, as it is used in many other applications such as manufacturing and building and even clothing. Perhaps we should also not be producing so much plastic, but oil-based materials will still be in demand. Lightweight carbon fibres in particular will be required. Perhaps governments can pay oil companies to clean up the environment as a way of bailing them out? Perhaps instead of looking for oil, they could use their knowledge and expertise to provide water in areas of drought? If we are to use water instead of oil and gas, then there will be a new requirement for infrastructure to get it to where it is needed. Already there are issues with providing fresh water to farms to meet the rising demand of food production.

Is there enough water?

One thing is for sure, fuel shortages will be a thing of the past. Just imagine ships and boats that can use nonpolluting water as fuel. They'll never run out wherever they are! Imagine also, water pumps, small and large, for a range of uses from agriculture, to flood zones and even getting drinking water to areas of drought.

Farm machinery can use local sources, even collected rainwater (a possible spin-off is to use animal urine but that's another story). Households can collect rainwater too. In dry areas, desalination processes will benefit from water power and if water has to be transported that can be done far cheaper too. There would be massive cost savings for industry too, especially with big energy users. Currently, in the UK (2022) energy providers are even going out of business and householders really suffering financially as energy prices soar.

Oil-producing countries may be very concerned about my revelations and proposals, but, there are massive benefits for them too. Think particularly about the Middle Eastern oil-rich nations, which previously benefited greatly from oil revenue and could save a fortune on water projects such as desalination and irrigation. They could easily turn their part of the desert green, boost tourism and never have to worry about supplies of freshwater, not to mention the requirement for expensive air conditioning, essential in extreme climates.

Pumping water up from underground aquifers would become feasible if the cost of running the pumps was reduced by using hydrogen from the very water that is being processed. Using satellite technology, scientists have discovered huge reservoirs of groundwater under the Sahara[1], amounting to 100 times the amount on the surface but, it's currently too expensive to recover.

All this new infrastructure brings many new opportunities and benefits for individuals and governments alike.

Book two?

We are in the midst of a climate emergency. Every time I turn on the TV there are numerous reports about the environmental crisis that we are in. Many questions are being asked and reports of the damage being done and lives being lost as a result of extreme weather events such as flooding and wildfires, are all too common. I cringe and sit uncomfortably watching, knowing that I am sitting on what I am convinced is the answer, a zero carbon, relatively simple technique to get clean-burning hydrogen from abundant water. It does hold the promise of powering our future where other efforts fall short.

Current hydrogen solutions are simply not the answer. Most require the dangerous storage of gas, being derived from carbon-rich coal gas, or needing expensive and complex fuel cells, which are simply not practical for the mass market. There currently seems to be a massive campaign to promote electric vehicles, but they need to source their power from somewhere else too. The infrastructure isn't there yet and it's mostly not green either, usually from coal-fired or natural gas power stations or extremely dangerous nuclear power plants.

For all these reasons, I have decided to get this book out as soon as possible and follow on with a sequel for everything else that I have researched but have not yet included. This decision, in part, is because there is also the HYDOD (Hydrogen On-Demand) website (www.hydod.com), which is much easier to populate with information and updates. I also wanted to include some experimental data of my own to support my claims and those of all the innovators that I have introduced here. However, I have had to focus on this edition and all my time has been taken up with writing, illustrating and publishing tasks. What I will do is promise to include those results in book two. They will also be part of a scientific report that I will submit to a respected scientific journal for proper peer review. (All backed up in the blockchain of web3...including the draft for book 2).

As a matter of urgency, I decided to get this book out as quickly as possible and also to as wide an audience as I can. Splitting the intended project into two editions in the first instance shortens the timescale required to get the message out. My intention is to make book two a much more practical guide, as well as having much more technical detail.

The idea is to bring this incredible technology out into the world as a solution to the current climate crisis. If having read this edition you are still not convinced, then there will be lots more information planned for the website and social media, especially the YouTube channel.

Along with the in-depth work on book two, my next stage is a lot of practical work leading to talks, demonstrations and presentations. Funds from sales of the book will be extremely helpful to fund subsequent stages of my mission. (Again, anyone considering preventing me from continuing my work is already too late, as, along with global book distribution, I have lodged all my research material from many years of study (including all my notes for book two), with various individuals, institutions and my legal team. Any attempt to do so will only result in a doubling of efforts to promote the whole technology).

How you can help.

Firstly you have already helped by buying this book, for which I thank you. The proceeds will go towards completing my mission to help achieve zero carbon with a clean healthy environment for everyone. Secondly, I need to get the message out to as many people as possible as soon as possible so as to achieve a critical mass of followers. At which point, governments or anyone with a vested interest in fossil fuels will no longer be able to silence those, like me, who attempt to adopt this clean, zero-carbon fuel source that is in abundance on our precious planet. I realise that the great majority of readers won't have the means or inclination to modify engines or otherwise introduce the technology outlined herein into their lives, but the main thing is for the reader to have some understanding of how radically it can help the looming environmental crisis. I really don't see any other real solutions. This will hopefully bring new opportunities to everyone and solutions to a few different problems. Does this sound like something that needs a closer look? About time we look at it seriously, I say. Many many thanks for purchasing this book. Don't forget to visit the website, social media accounts and my

YouTube channel. The proceeds of this book will also help me proceed with further practical research and promotion of the ideas that I have covered. Everything I do is now open-source, so let's now all work together for a better world rather than greed.

Summary:

We have looked at hydrogen from water on-demand as being a far better solution to all our energy needs than existing alternatives. Land is too precious to cover in vast swathes of solar panels, which only work during daylight hours. Wind power also relies on fickle weather and offshore wind and wave plants are subject to the extremes of the marine environment, which also cost a fortune in maintenance. Existing hydrogen and hybrid systems have logistical problems too, as well as storage and transportation issues. The 'electric' revolution currently underway has infrastructure problems as well (the different types of connectors and location network), also range issues. At the moment, most are actually powered at source, by coal-fired power stations! Hydrogen sourced from water is clean, green energy, abundant and safe to use in this way. It's pollution-free, carbon-free and so easy to implement that hobbyists are already producing promising results in their sheds, garages and basements with simple off-the-shelf parts. I think you can see that this really is the fuel of the future, and there it was, hiding in plain sight, in the water that surrounds us.

And what of the poor oil companies? What will become of them now that there's a viable zero-carbon solution? I imagine that they have made enough profit to last well into the next century, however, if water is to be used as 'the new fuel' it must be managed properly. Freshwater is already in short supply and needs to be protected. The knowledge and skills that oil workers have can be put to finding underground aquifers, creating water pipelines to distribute it to where it's needed (for example supplying seawater to inland areas) and desalination plants to create more fresh water in places of greatest shortage. Food production is also in need of vast quantities of fresh water and there are already problems with supply and distribution within that sector. If you ask me, there are many opportunities for new careers in the management of water, especially if the world turns to using it for fuel as well as all the existing requirements. This technology has the potential to feed the world as well as provide fresh water to where it's needed, but mainly to massively reduce the cost of energy (it still needs infrastructure), and clean up the planet in the process.

Warning...

Please do not attempt any of the procedures mentioned in this book unless you are qualified and completely competent in the disciplines being discussed.

Even if you are very sure about what you are doing, electricity can be deadly, hydrogen explosive and even water turned poisonous while using the methods outlined.

Electrical circuits can still give a lethal shock even when turned off and components like capacitors can contain a residual charge even when disconnected from the power supply. Extremely high voltages can be generated by coils and electricity, being invisible, can take even the most experienced experimenter by surprise. Water can also be turned toxic during the process of electrolysis, for example, when using seawater, poisonous chlorine gas can be given off. Hydrogen is of course highly volatile and must not be allowed to build up in small spaces. Even safety devices like flash-back arrestors should only be used once and the correct type for hydrogen employed, as they may not function correctly if they were designed for say, oxygen-acetylene as used in welding. Mistakes can be made even by the most safety-conscious of us. Wires can touch unexpectedly, chemicals can be spilt and errors made in electronic or electrical circuits can lead to unexpected results, such as fire due to overheating. With care, all this technology can be implemented safely, especially when producing hydrogen on-demand as no gas needs to be stored. However, I urge you to proceed with great caution.

About the author

Paul Adams was born in North Cheshire, United Kingdom in 1955 and went to school in Manchester, UK. Influenced by his father, who was a sign writer and artist as well as his grandfather, who was a pioneer in photography, he attended Manchester High School of Art. This resulted in a solid grounding in art, design and photography, though the main theme of his career has been in publishing, specifically technical publishing.

Whilst still at school he was commissioned by a school history teacher to produce illustrations for an archaeological research project at Manchester University (in the UK and on location in Benghazi, Libya), which led to his first paid illustration work intended for publication. Following this path briefly led to study at Bradford University where he was introduced to computing which henceforward became a passion. This came to the fore when software had advanced to the stage where most commercial graphics and publishing were produced using computer software.

Having had early experience producing accurate technical renderings of archaeological finds for publication coupled with a fascination for all things technical, led to a career in technical publications. Working for technical publications agencies in Bradford and Leeds, Manchester, Bristol and London he built up experience across many areas of industry and commerce.

More recently after working briefly as an assistant publisher at the University of Bath, he changed career direction and became an instructor of computer software and ultimately became an Adobe Certified Instructor in their 'Creative Suite', the industry standard for publishing and graphics. Subsequently studying for a science degree with the Open University, research is one skill that has followed through, possibly due to a great love of libraries, lateral thinking skills and most recently helped by modern internet resources.

He is well-travelled, having relatives in Houston, Atlanta, New York and Calgary, USA and besides the USA has lived for periods of time in places across Europe and further afield such as India. He has a passion for sports aviation and a love of technology. Although he has officially reached retirement age, he is showing no signs of stopping.

GLOBAL EMERGENCY

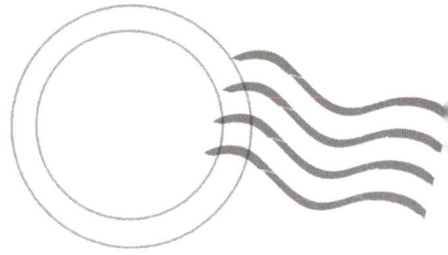

An open letter to world governments and global oil corporations

A lot of money has been made from oil revenue and the tax it generates over the past century but at <u>great cost to the health of your citizens and the environment</u>. It has also been the cause of many costly conflicts around the world. Costly in terms of money and people and resources.

Oil is a finite resource that is fast running out and it doesn't make sense to burn it (at the rate of 100 million barrels per day when we need it for many other important roles such as the production of modern materials (such as oils, fertilisers, polymers, graphene and asphalt).

The world's climate has now reached a crisis point, even to the stage that it threatens the continuation of our human species and all the fauna and flora that we depend on. Climate change has reached a crisis point, including devastating floods and wildfires across the globe, notwithstanding the health effects of continuing atmospheric pollution.

I hope this book will now convince your top scientists to look again at the possibility of using water as a source of hydrogen on-demand and the huge benefits that it opens up. We do not want to hear of this technology being suppressed any longer with its forward-thinking innovators being bribed, gagged, or even murdered (as outlined throughout this book).

By the time you read this, my book will now be published and hopefully selling around the world, so the message will be out. Please don't shoot the messenger! I am not the one who made these discoveries and wrote the patents herein disclosed. I am just a researcher who gathered the evidence from freely available data available online. (and I will be freely sharing my research, by open-source and blockchain methods, so that it will be very resilient to government control or hacking attempts).

I see it as my mission as the catalyst that brings together all the people who believe in this technology and the scientific community who can make it work with our future fuel requirements. Believe me, this will benefit everyone. Environmentally friendly, carbon-free energy, at a massively reduced cost, so fresh water from efficient desalination plants, and underground aquifers, better food production through improved irrigation, with cleaner air and water supplies.

There is still plenty of money to be made (and saved, through a reduction in environmental disasters, health issues, transportation costs and military spending) because this new technology needs new infrastructure, new machines, new water pipelines and new desalination plants, etc. And, of course, we still need some oil refining to produce many other modern materials.

So I urge you. Get your top scientific and engineering departments to look again at this technology and stop trying to convince people otherwise. Hydrogen and Oxygen *can* be *efficiently* released from plain old water by using a perfectly tuned resonant frequency on a high voltage, low current, AC supply, as I have now revealed!

Paul Adams

This is not just a book, it is part of a project to bring what I have discovered to life. There are many experimenters out there (evidenced by the likes of YouTube, Pinterest and Instagram), who are achieving great results with these methods, as well as pioneers whose work has been lost or suppressed. One essential thing that no one has yet done, to my knowledge, is to submit the findings to a scientific journal for peer review. That is my mission. This needs to be done in the correct way, following certain procedures and protocols, which is a very expensive process for which I need your help. Even a small donation will help. In return, I'm making everything 'Open-Source', in other words sharing all my results openly, to hopefully rapidly propagate this clean, environmentally friendly technology around the world before it is too late. Also, sharing everything means that if I'm targeted, like others before me, it can't prevent the ideas from getting out this time. The profits from the sale of this book will also help to provide equipment and resources as well as funding for advertising and promotion. Thank you for your interest and support.

Website for information: **www.hydod.com**

PayPal donations to email@hydod.com

Crowd-funding: https://ko-fi.com/hydod

Web3 blockchain address HYDODproject.blockchain

Many thanks for purchasing and reading this book. I know that by its very nature it is unavoidably technical, but I hope that has not put off the readers who are after all from many different backgrounds and levels of education. What really matters is that you got something interesting and relevant to your level of understanding. I hope to achieve a critical mass where this technology is taken on board by those who have the power to make it happen and replace our overuse of expensive and dangerous fossil fuels.

Appendix 1

For those of you who, like me, have a unquenchable thirst for technical information, I thought that I had better include these interesting articles at the end of the book. Firstly I have a reference to the article in the Journal of Plasma Physics which proves that you can get more energy out of a chemical process to fracture water than you put in. Secondly there is a reference to the news that the US Navy used seawater to provide fuel whilst the government was denying the possibility. Then I have relevant extracts from Andrija Puharić's Patent US 4,394,230 'Method and apparatus for splitting water molecules' (although I have given you links to the full patent too). Then Puharić's detailed article 'Acoustic Decomposition of Water by the Phonon Effect: Towards a Viable Clean Fuel System'. Next the peer-reviewed article by American physicist/parapsychologist Dr Elizabeth Rauscher (1937-2019). 'Water-splitting system for cars' about Puharić's work. Lastly, I have a note about the project's Web3-Decentralised Website & Filestore and a book recommendation for those who want to know more about using hydrogen in a practical way.

Proof that energy output can exceed electrical input!

Often people say (and truly believe) that it's impossible to get more energy out of water than you 'put in' to the process of releasing it. But that simply isn't true and has been scientifically proven.

Arc-liberated chemical energy exceeds electrical input energy

Journal of Plasma Physics' Volume 63 Issue 2
Published online by Cambridge University Press: 01 February 2000

This paper reports the first experimental results in which the kinetic energy of cold fog, generated in a water arc plasma, exceeds the electrical energy supplied to form and maintain the arc. The cold fog explosion is produced by breaking down a small quantity of liquid water and passing a kiloampere current pulse through the plasma. The 90-year history of unusually strong water arc explosions is reviewed. Experimental observations leave little doubt that internal water energy is being liberated by the sudden electrodynamic conversion of about one-third of the water to dense fog. High-speed photography reveals that the fog expels itself from the water at supersonic velocities. The loss of intermolecular bond energy in the conversion from liquid to fog must be the source of the explosion energy.

"..energy is that stored by hydrogen bonds between the water molecules."

... To dissociate this amount of water into oxygen and hydrogen would require 10 kJ of energy. Hence the fog explosion is unlikely to be caused by electrolytic dissociation of water molecules. Without this dissociation, the most likely source of the explosion energy is that stored by hydrogen bonds between the water molecules.

PETER GRANEAU, NEAL GRANEAU & GEORGE HATHAWAY

1. https://bit.ly/3KRaXhH

US Naval Research Lab turned seawater into fuel.

Ships burn a huge amount of fuel, often as much as 1,000 gallons an hour and of course, they have to find somewhere to dock and refuel. Naval fleets often use tankers to transfer fuel, which can also be problematic in rough weather. Plus they need fuel for their fighter jets on board aircraft carriers. But look at all that water they are sitting in. It contains hydrogen after all. So the US Navy had a brilliant idea (I wonder where they got that?). Why not turn the seawater into fuel? So, a U.S. Naval Research Laboratory proved that it is in fact possible to power the ship's engines with fuel from seawater[1]. A proof-of-concept test was performed using what they called a 'proprietary electrochemical device', which is in fact a specialized catalytic converter used to recover carbon dioxide from the seawater, with hydrogen as a by-product. They then bounced the two gases off each other to manufacture liquid hydrocarbon fuel. The resulting fuel can simply be used in the ships' existing engines.

The 'fact' that water isn't an economically viable proposition is a **perpetuating myth** that needs exposing. NASA uses it, the military use it and yet they tell us that we must keep burning fossil fuels. It doesn't make sense.

The Pentagon knew of Andrija Puharić's work on water-fuel technology, leading him to worry about the patent being made secret in order to fuel their stealth submarines (see chapter 10, page 134). When he died, the Planetary Association for Clean Energy (PACE) received a request from Rolls Royce aircraft engines in Atlanta as they were considering using his system for their future jet engines. NASA too held meetings with Stanley Meyer about his inventions (oxygen and hydrogen from water being very useful in space).

---

1. US Navy turns seawater into jet fuel
   https://bit.ly/38IhDlu
   https://www.youtube.com/watch?v=Fcc-cTCVY64
   US Navy Develops 'Game-Changer' Technology To Turn Seawater Into Fuel
   https://bit.ly/3zbdzoE
   How Ships Could Produce an Unlimited Amount of Their Own Fuel
   https://bit.ly/3zgx4fD

Extracts from Patent US 4,394,230
Andrija Puharić 1983-07-19

METHOD AND APPARATUS FOR SPLITTING WATER MOLECULES

Disclosed herein is a new and improved thermodynamic device to produce hydrogen gas and oxygen gas from ordinary water molecules or from seawater at normal temperatures and pressure. Also disclosed is a new and improved method for electrically treating water molecules to decompose them into hydrogen gas and oxygen gas at efficiency levels ranging between approximately 80-100%. The evolved hydrogen gas may be used as a fuel; and the evolved oxygen gas may be used as an oxidant.
(https://patents.google.com/patent/US4394230A/en?oq=US+4394230)

BACKGROUND OF THE INVENTION

The scientific community has long realized that water is an enormous natural energy resource, indeed an inexhaustible source, since there are over 300 million cubic miles of water on the earth's surface, all of it a potential source of hydrogen for use as fuel. In fact, more than 100 years ago Jules Verne prophesied that water eventually would be employed as a fuel and that the hydrogen and oxygen which constitute it would furnish an inexhaustible source of heat and light. Water has been split into its constituent elements of hydrogen and oxygen by electrolytic methods, which have been extremely inefficient, by thermochemical extraction processes called thermochemical water-splitting, which have likewise been inefficient and have also been inordinately expensive, and by other processes including some employing solar energy. In addition, artificial chloroplasts imitating the natural process of photosynthesis have been used to separate hydrogen from water utilizing complicated membranes and sophisticated artificial catalysts. However, these artificial chloroplasts have yet to produce hydrogen at an efficient and economical rate.

These and other proposed water splitting techniques are all part of a massive effort by the scientific community to find a plentiful, clean, and inexpensive source of fuel. While none of the methods has yet proved to be commercially feasible, they all share in common the known acceptability of hydrogen gas as a clean fuel, one that can be transmitted easily and economically over long distances and one which when burned forms water.

DESCRIPTION OF INVENTION (Short)

Section 1-Apparatus of Invention

The apparatus of the invention consists of three components, the electrical function generator, the thermodynamic device, and the water cell.

COMPONENT I. The Electrical Function Generator

This device has an output consisting of an audio frequency (range 20 to 200 Hz) amplitude modulation of a carrier wave (range 200 Hz to 100,000 Hz). The impedance of this output signal is continuously being matched to the load which is the second component, the thermodynamic device.

COMPONENT II. The Thermodynamic Device

The thermodynamic device is fabricated of metals and ceramic in the geometric form of a coaxial cylinder made up of a centered hollow tubular electrode which is surrounded by a larger tubular steel cylinder, said two electrodes comprising the coaxial electrode system which forms the load of the output of the electrical function generator, Component I. the coaxial electrode system which forms the load of the output of the electrical function generator, Component I.

COMPONENT III. The Water Cell

The water cell is a part of the upper end of Component II and has been described. Component III consists of the water and glass tubes contained in the geometrical form of the walls of the cell in Component II, the thermodynamic device.

The elements of a practical device for the practice of the invention will include:

(A) Water reservoir; and salt reservoir; and/or salt

(B) Water injection system with a microprocessor or other controls that sense and regulate (in accordance with the parameters set forth hereinafter):

a. carrier frequency

b. current

c. voltage

d. RC relaxation time constant of water in the cell

e. nuclear magnetic relaxation constant of water

f. temperature of hydrogen combustion

g. carrier waveform

h. RPM of an internal combustion engine (if used)

i. ignition control system

j. the temperature of region to be heated;

(C) An electrical ignition system to ignite the evolved hydrogen gas fuel.

SUMMARY OF THE PRESENT INVENTION (Note: Author's emphasis)

In classical quantum physical chemistry, the water molecule has two basic bond angles, one angle being 104.45°, and the other angle being 109.5°.

The present invention involves a method by which a water molecule can be energized by electrical means so as to shift the bond angle from the 104.45° configuration to the 109°28' tetrahedral geometrical configuration. An electrical function generator (Component 1) is used to produce **complex electrical wave form frequencies which are applied to, and match the complex resonant frequencies of the tetrahedral geometrical form of Water.**

It is this complex electrical wave form applied to water which is contained in a special thermodynamic device (Component II) which **shatters the water molecule by resonance into its component molecules-hydrogen an oxygen.**

The hydrogen, in gas form, may then be used as fuel; and oxygen, in gas form is used as an oxidant. For example, the thermodynamic device of the present invention may be used as a hydrogen fuel source **for any existing heat engine-such as, internal combustion engines of all types, turbines, fuel cells, space heaters, water heaters, heat exchange systems, and other such devices. It can also be used for the de-salinization of seawater, and other water purification purposes.** It can also be applied to the development of new closed-cycle heat engines where water goes in as fuel, and water comes out as a clean exhaust.

Generation of the complex wave form frequencies from Component I to match the complex wave form resonant frequencies of the energized and highly polarized water molecule in tetrahedral form with angles, 109°28' are carried out in Stage C.

In the operation of the invention active bubble electrolysis of water is initiated following Stage B, phase 3 by setting (automatically) the output of Component I to:

I = 1mA., E=22VAC-rms,

causing the rippled square wave pulses to disappear with the appearance of a rippled sawtooth wave. The basic frequency of the carrier now becomes, f_c=3980 HZ.

The wave form now automatically shifts to a form found to be the prime characteristic necessary for optimum efficiency in the electrolysis of water and illustrated in FIG, 11. In the waveform of FIG. 11, the fundamental carrier frequency, f_c=3980 Hz., and a harmonic modulation of the carrier is as follows:

1st Order Harmonic Modulation (OHM)=7960 Hz.

2nd Order Harmonic Modulation (II OHM)= 15,920 HZ.

3rd Order Harmonic Modulation (III OHM)=31,840 HZ.

4th Order Harmonic Modulation (IV OHM)=63,690 HZ.

*What is believed to be happening in this IV OHM effect is that **each of the four apices of the tetrahedron water molecule is resonant to one of the four harmonics observed**. It is believed that the combination of negative repulsive forces at the outer electrode with the resonant frequencies just described work together to shatter the water molecule into its component hydrogen and oxygen atoms (as gases).*

FIG. 1

COMPONENT I
SIGNAL GENERATOR
BLOCK DIAGRAM

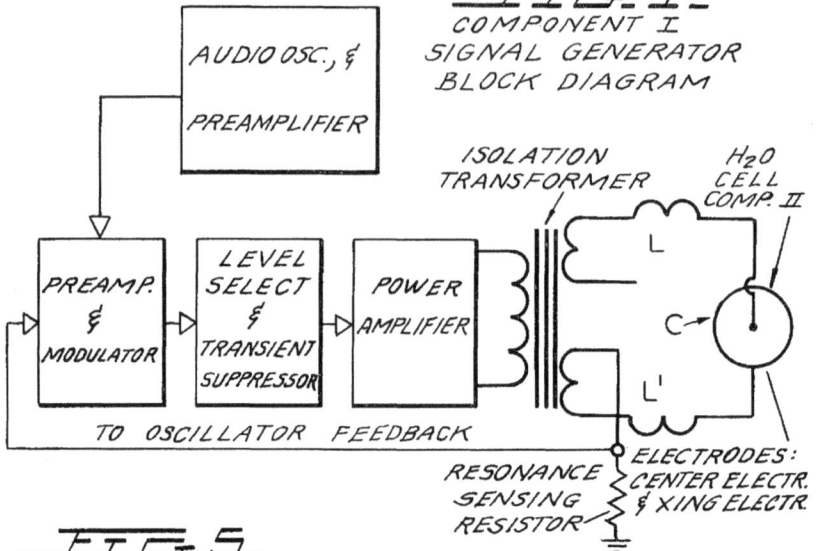

AUDIO OSC., &
PREAMPLIFIER

ISOLATION
TRANSFORMER

H_2O
CELL
COMP. II

PREAMP.
&
MODULATOR

LEVEL
SELECT
&
TRANSIENT
SUPPRESSOR

POWER
AMPLIFIER

L

C

L'

TO OSCILLATOR   FEEDBACK

RESONANCE
SENSING
RESISTOR

ELECTRODES:
CENTER ELECTR.
& XING ELECTR.

FIG. 5

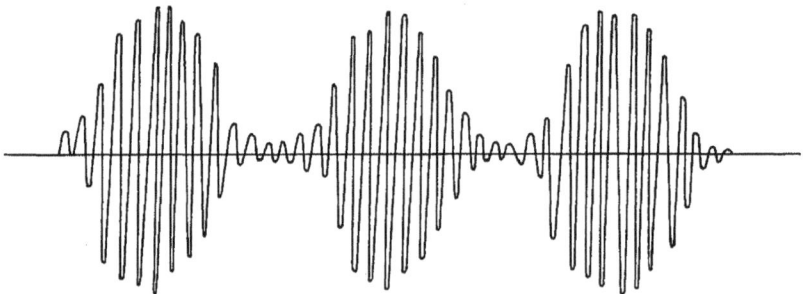

AMPLITUDE MODULATED 90° CARRIER SINE WAVE

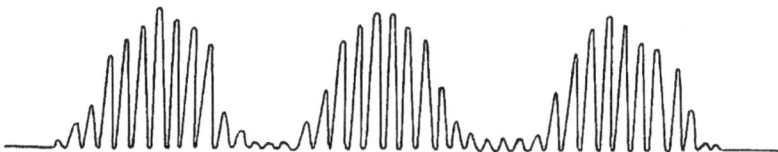

HALF-WAVE VECTIFICATION OF ABOVE SIGNAL

Acoustic Decomposition of Water by the Phonon Effect: Towards a Viable Clean Fuel System

Cutting the Gordian knot of the great energy bind

by Andrija Puharić

(reproduced in full by kind permission of the Planetary Association for Clean Energy, Inc. Ottawa, Ontario. Canada).

From the Ancient Greek legend of Phrygian Gordium, "Cutting the Gordian knot" is a way of slicing through a seemingly insurmountable problem (such as trying to untie an impossibly tangled knot). Perhaps there's a much easier solution than at first perceived.

From First International Symposium on non-conventional Energy Technology (PACE) First Published in the Planetary Association for Clean Energy Newsletter December 1981

Visionary scientists tell us that the ideal fuel in the future will be as cheap as water; that it will be non-toxic both in its short term, and in its long term effects; that it will be renewable in that it can be used over and over again; that it will be safe to handle, and present minimal storage and transportation problems and costs. And finally that it will be universally available anywhere on earth.

The fuel is water. It can be used in its fresh water, salt water, brackish, and in its snow-and-ice forms. When such water is decomposed by electrolytic fission into hydrogen and oxygen gases - it becomes a high-energy fuel with three times the energy output which is available from an equivalent weight of high-grade gasoline.

Then why is water not being used as a fuel? It costs too much with existing technology to convert water into two gases.

The cost of producing hydrogen is directly related to the cost of producing electricity. Hydrogen as produced today is generally a by-product of off-peak-hour electrical production in either nuclear or hydroelectric plants, the cheapest way of making hydrogen and costs about 21% more than electricity to produce.

The requisite breakthroughs in Hydrogen technology in the opinion of experts, if the hydrogen production cost component of its total cost could be reduced three-fold, it would become a viable alternative energy source. In order to achieve such a three-fold reduction in production cost, several major breakthroughs would have to occur:

1. **Endergonic reaction, supra.** A technological breakthrough that permits 100% conversion efficiency of water by electrolysis fission into the two gases, Hydrogen as fuel, and Oxygen as oxidant.

2. **Hydrogen Production in situ.** A technological breakthrough that eliminates the need and cost of hydrogen liquefaction and storage, transmission, and distribution, by producing the fuel **in situ**, when and where it is needed.

3. **Exergonic Reaction, supra.** A technological breakthrough which yields a 100% efficient energy release from the combination of hydrogen and oxygen into water in an engine that can utilize the heat, or steam, or electricity thus produced.

4. **Engine Efficiency.** By a combination of the breakthroughs outlined above, 1), 2), and 3) utilized in a highly efficient engine to do work, it is possible to achieve a 15% to 20% surplus of energy return over the energy input, theoretically.

An answer

A Thermodynamic Device has been invented which produces hydrogen as fuel, and oxygen as an oxidant, from ordinary water, or from seawater, when needed, or where needed, for energy production, thereby eliminating the cost and hazard of liquefaction, storage, transmission, and distribution. The savings of this aspect of the invention alone reduces the total cost of hydrogen by about 25%

This Thermodynamic Device is based on a new discovery: the efficient electrolytic fission of water into hydrogen gas and oxygen gas by the use of low-frequency alternating currents as opposed to the conventional use of direct current, or ultra-high frequency current, today. Such gas production from water by electrolytic fission approaches 100% efficiency (under laboratory conditions and measurements). No laws of physics are violated in this process.

This Thermodynamic Device has already been tested at ambient pressures and temperatures from sea level to an altitude of 10,000 feet (3,000 Metres) without any loss of its peak efficiency. The device produces two types of gas bubbles; one type of bubble contains hydrogen gas, the other type contains oxygen gas. The two gases are thereafter easily separable by passive membrane filters to yield pure hydrogen gas and pure oxygen gas.

The separate gases are now ready to be combined in a chemical fusion with a small activation energy such as that from a catalyst or an electrical spark, and yield energy in the form of heat, or steam, or electricity - as needed. When the energy is released by the chemical fusion of hydrogen and oxygen, the exhaust product is clean water. This water exhaust can be released into nature, and then renewed in its energy content by natural processes of evaporations, solar irradiation in cloud form, and subsequent precipitation as rain on land or sea, and then collected again as a fuel source. Or, the exhaust water can have its energy content pumped up by artificial processes such as through solar energy acting through photocells. Hence, the exhaust product is both clean and renewable.

The fuel hydrogen, and the oxidant oxygen can be used in any form of heat engine as an energy source - *if economy is not an important factor*. But the practical considerations of maximum efficiency dictate that a low-temperature Fuel Cell with its direct chemical fusion conversion from gases to electricity offers the greatest economy and efficiency for small power plants (less than 5 kilowatts).

For large power plants, steam and gas turbines are the ideal heat engines for economy and efficiency. ***With the proper engineering effort, automobiles could be converted rather easily to use water as the main fuel source*.**"

The Thermodynamic Device (TD) is made up of three principal components:

Component I The Electrical function Generator. (See Figure 1).

This electronic device has a complex alternating current output consisting of an audio frequency (range - 20 to 200 Hz) amplitude modulation of a carrier wave (range - 200 to 100,000 Hz).

The output is connected by two wires to Component II at the center electrode, and at the ring electrode. The impedance of this output signal is continuously being matched to the load which is the water solution in Component II.

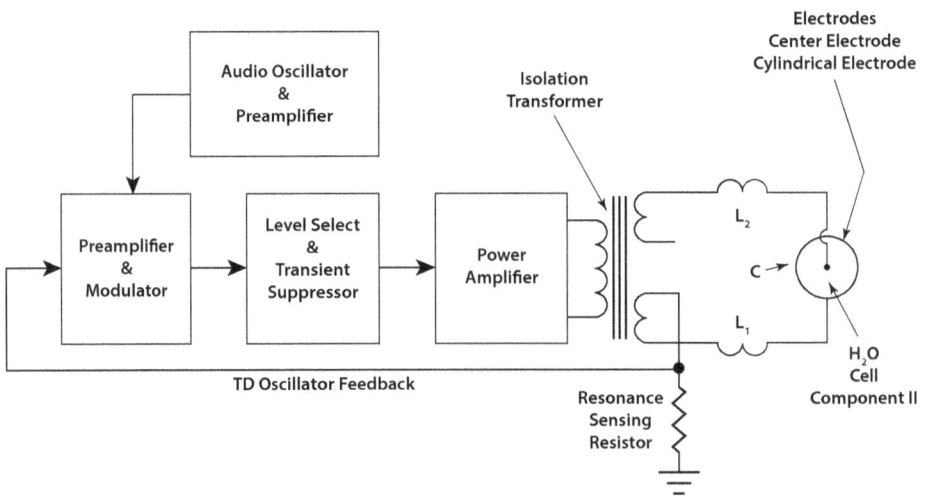

Figure 1. COMPONENT I SIGNAL GENERATOR BLOCK DIAGRAM

Component II The Thermodynamic Device. (See Figure 2).

The TD is fabricated of metals and ceramic in the geometric form of a coaxial cylinder made up of a centred hollow tubular electrode which is surrounded by a larger tubular steel cylinder.

These two electrodes comprise the coaxial electrode system energized by Component I. The space between the two electrodes is, properly speaking, Component III which contains the water solution to be electrolysed. The centre hollow tubular electrode carries water into the cell and is further separated from the outer cylindrical electrode by a **porous ceramic vitreous material**. The space between the two electrodes contains two lengths of tubular Pyrex glass, shown in Figures 2 and 3. The metal electrode surfaces in contact with the water solution are coated with a nickel alloy.

Component III. The weak electrolyte water solution. (Figure 3).

This consists of the water solution, the two glass tubes, and the geometry of the containing wall of Component II. It is the true load for Component I, and its electrodes of Component II. The Component III water solution is more properly speaking, ideally a 0.1540 Molal Sodium Chloride solution, and as such is a weak electrolyte. In Figure 4 is the hypothetical tetrahedral structure of water molecule, probably in the form in which the complex electromagnetic waves of Component I "see" it. The centre of mass of this tetrahedral form is the oxygen atom. The geometric arrangement of the p electrons of oxygen probably determine the vectors, $i(L_1)$ and $i(L_2)$ and $i(H_1)$ and $i(H_2)$ which in turn probably determine the tetrahedral architecture of the water molecule. The p electron configuration of oxygen is shown in Figure 5. Reference to Figure 4 shows that the diagonal of the right side of the cube has at its corner terminations the positive charge hydrogen (H^+) atoms; and that the left side of the cube diagonal has at its corners the lone pair electrons, ($e-$). It is to be further noted that this diagonal pair has an orthonormal relationship.

EM-Water molecule interactions

The complex electromagnetic wave may be portrayed as the tetrahedral water molecule "sees" it. The first effect "felt" by the water molecule is in the protons of the vectors; $i(H_1)$ and $i(H_2)$. These protons "feel" the 3-second cycling of the amplitude of the carrier frequency and its associated side bands as generated by Component I. This sets up a rotation

263

moment of the proton magnetic moment which one can clearly see on the XY plot of an oscilloscope, as a hysteresis loop figure. *This hysteresis loop does not appear in the liquid water sample until all the parameters of the three components have been adjusted to the configuration.* The hysteresis loop gives a vivid portrayal of the *nuclear magnetic relaxation cycle of the proton* in water.

The next effect felt by the water molecule is the Component I carrier resonant frequency, f_o. At the peak efficiency for electrolysis, the value of f_o is 600 Hz + 5 Hz. This resonance, however, is achieved through the control of two other factors. The first is the molal concentration of salt in the water. This is controlled by measuring the conductivity of the water through the built-in current meter of Component I. There is maintained an ideal ratio of current to voltage, $I/E = 0.01870$ which is an index to the optimum salt concentration of 0.1540 Molal.

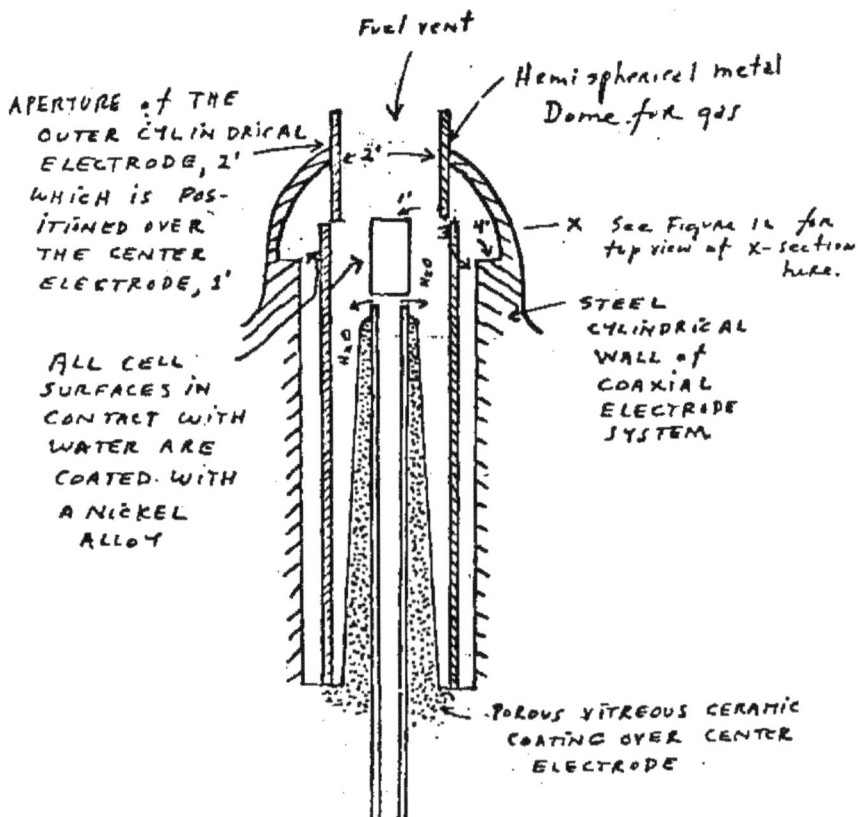

Figure 3. COMPONENT III THE WATER CELL SECTION

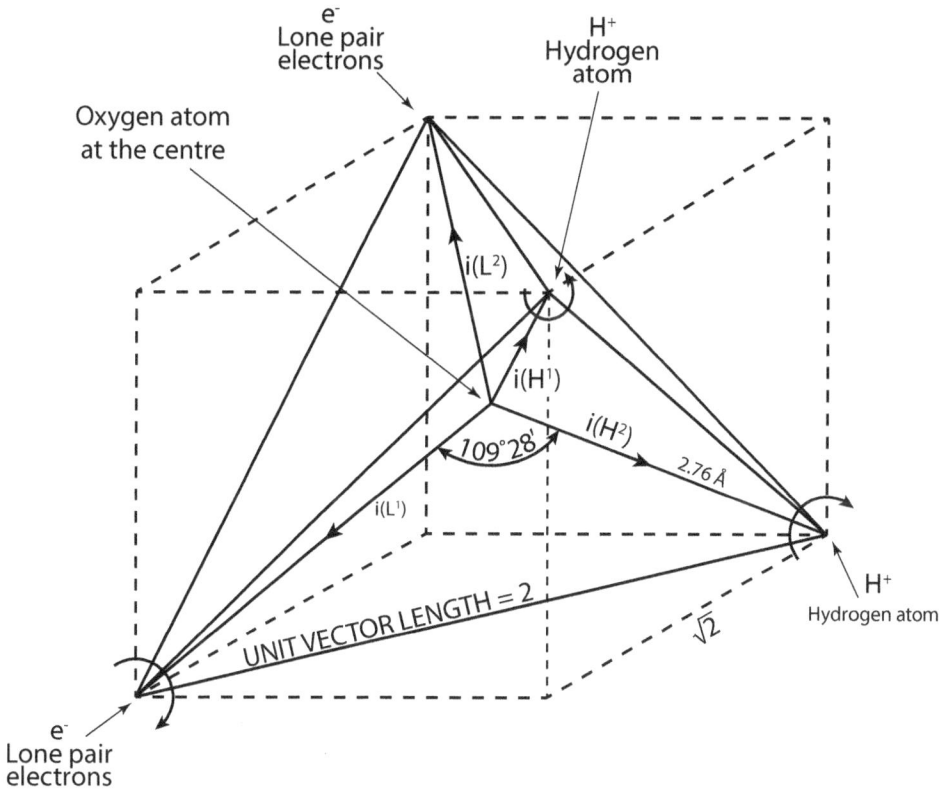

Figure 4. THE WATER MOLECULE IN TETRAHEDRAL FORM

Hydrogen bonding occurs only along the four vectors pointing to the four vertices of a regular tetrahedron, and in the above drawing, we show the four unit vectors along these directions originating from the oxygen atom at the centre. $i(H_1)$ and $i(H_2)$ are the vectors of the hydrogen bonds formed by the molecule i as a donor molecule. $i(L_1)$ and $i(L_2)$ denote the unit vectors along the direction of the bonds formed by molecule i, as an acceptor molecule. These are assigned to the lone pair electrons. Molecules i are the neighbouring oxygen atoms at each vertex of the tetrahedron.

Electro thermodynamics

The second factor which helps to hold the resonant frequency at 600 Hz is the gap distance, between the centre electrode, and the ring electrode of Component II. This gap distance will vary depending on the size scale of Component II, but again the current flow, I, is used to set it to the optimal distance when the voltage reads between 2.30 (rms) to 2.46 (rms) volts, at resonance, f_0 and at molal concentration, 0.1540. The molal concentration of the water is thus seen to represent the electric term of the water molecule and hence its conductivity.

The amplitude modulation of the carrier gives rise to sidebands in the power spectrum of the carrier frequency distribution. *It is these sidebands which give rise to an acoustic vibration of the liquid water, and it is believed to the tetrahedral water molecule.*

The importance of the phonon effect - the acoustic vibration of water

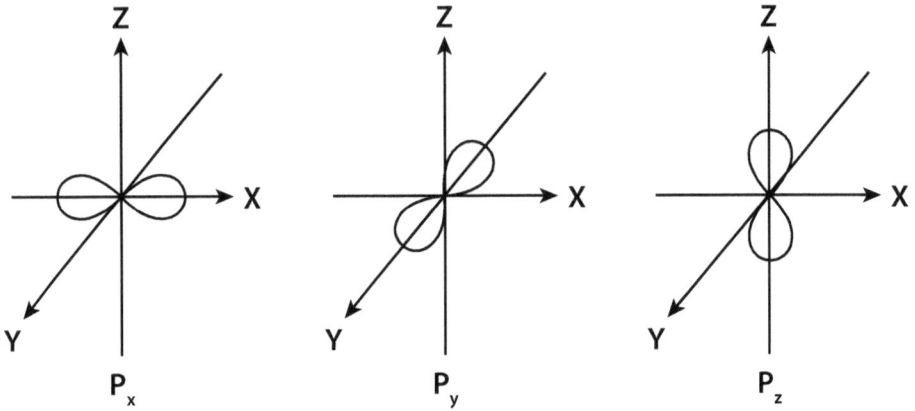

P_x P_y P_z

Electron Configuration

Element				
H	$1s^1$			
O	$1s^2$	$2s^2$	$2p^4$	

$(2p^1)_x$ $(2p^1)_y$ $(2p^1)_z$

Arrows indicate pairing of electrons in spin

Figure 5. ELECTRON ORBITALS

in electrolysis was discovered in a roundabout way. Research work with Component I had earlier established that it could be used for the electrostimulation of hearing in humans. When the output of Component I is comprised of flat circular metal plates applied to the head of normal hearing humans, it was found that they could hear pure tones and speech simultaneously. Acoustic vibration could also be heard by an outside observer with a stethoscope placed near one of the electrodes on the skin. It was observed that the absolute threshold of hearing could be obtained at 0.16 mW (rms), and by calculation that there was an amplitude of displacement of the eardrum of the order of 10^{-11} meter, and

a corresponding amplitude of the cochlear basilar membrane of 10^{-13} meter. As a corollary to this finding, I was able to achieve the absolute reversible threshold of electrolysis at a power level of 0.16 mW (rms). By carrying out new calculations, I was able to show that the water solution was being vibrated with a displacement of the order of $1A = 10^{-10}$ meters. *This displacement is of the order of the diameter of the hydrogen atom.*
Thus it is possible that the acoustic phonons generated by audio sidebands of the carrier are able to vibrate particle structures within the unit water tetrahedron.

(Andrija Puharić)

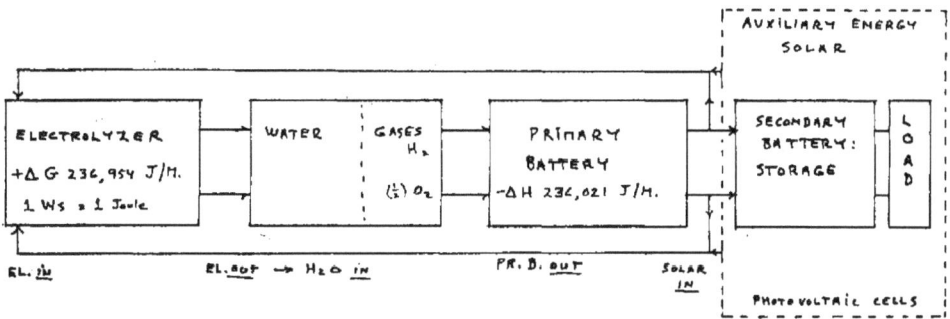

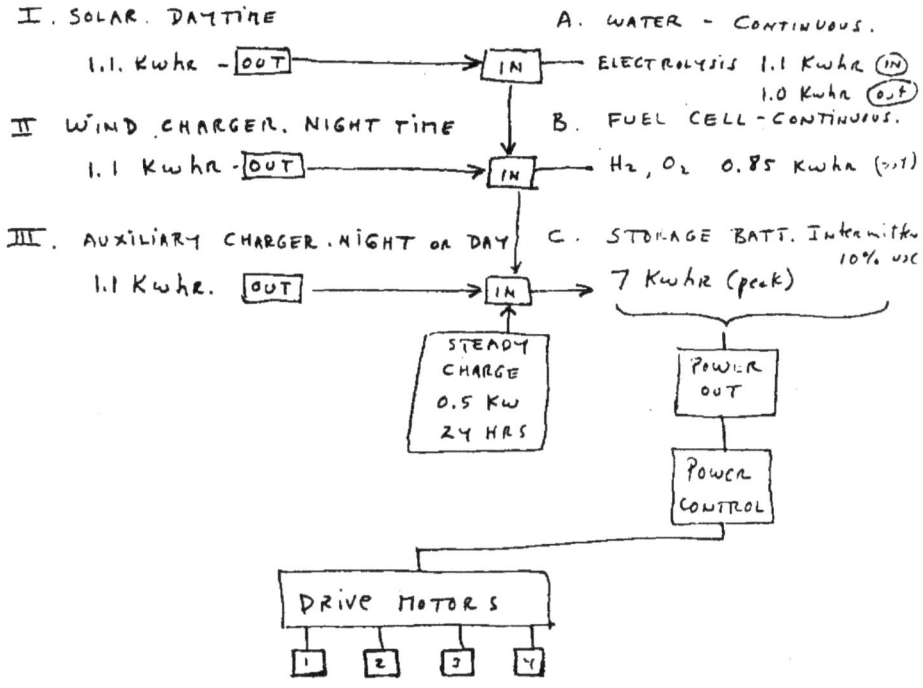

Puharić water-splitting system for cars

(By Dr Elizabeth Rauscher[2] (1937-2019), American physicist/parapsychologist In the Planetary Association for Clean Energy I Vol 8 (2) 14).

By 1980, after years of research, in part inspired by careful analysis of Nikola Tesla's understanding of electrical resonance, Andrija Puharić developed and perfected a method of splitting water molecules. Complex elliptical waveforms form resonant frequencies of tetrahedral water molecules.

These waveforms resonate water molecules and shatter them, thereby liberating hydrogen and oxygen. The system was first announced at two of the Association's functions: one, at the Royal Society of Canada hall in Ottawa (recorded in a video, *Energy from Water*) and the other, at the *First International Symposium for Non-conventional energy technology* (October 1981, University of Toronto), where it is described in detail in the proceedings.

The system was test-driven in his mobile home using water as an autonomous fuel system for several hundred thousand kilometres in several trips across North America, from Canada to Mexico. On one occasion, In the Sonoran desert, water from the salt flats was the required fuel to keep going; on another, snow from a high Mexican mountain pass made do.

On July 19, 1963, the system was credited with U.S. Patent 4,394,230, *Method and apparatus for splitting water molecules.*

In the Puharić electrolysis/hydrolysis system to produce fuel to power a vehicle, alternating current (AC) conversion to H_2 and O_2 occurs in the vehicle itself, recombined in a continuous flow system. The power supply for the AC system could be charged (by an external power source) DC batteries with DC to AC converters. Again, power loss needs to be examined. The creation and utilization of H_2 directly has the advantage that explosive tanks of H_2 are eliminated. The system's efficiency is a key issue here.

Enhancing car's electric drive

Essentially, the system enhances a car's electric power system, vastly increasing its range. One of the problems with direct battery-operated electric cars is the bulkiness of battery arrays and the need for frequent recharging. The battery system, in conjunction with the water system, allows a continuous process, minimizing storage of combustible materials, such as gasoline or H_2.

Another system utilizing water as the working substance is the steam engine, such as was used in the *Stanley Steamer*, which is unlikely to come back into use. In the "steamer", coal is required to create steam. Here, another source is needed, such as charged batteries to create H_2 to burn.

Efficiency tests: up to 110%

In April 1981, efficiency tests of the thermodynamic electrolysis/hydrolysis cells were conducted at **Gollob Analytical Laboratory**, New Jersey (100% efficiency defined as the number of mole electrons of current to completely disassociate a given fixed amount of water at a fixed temperature). In the 3 samples tested, the average apparent efficiency was 90%, with one at 100% +/- 10%, (based on 120% as defined in fuel cell analysis).

These analyses were based on the conventional view of thermodynamics. Ilya Prigogine has formulated other governing thermodynamic systems for open (flow), non-equilibrium, non-linear processes. [2] 100% efficiency 6.95 cm^3 H_2/minute ampoules at 0°C and 1-atmosphere pressure.

When $H_2(g)$ and $O_2(g)$ are generated by electrolysis, the electrolytic cell absorbs heat from the surrounding environment to remain isothermal. The system is therefore open. Electrical, mechanical and biological systems have been designed to examine non-linear, coherent resonant phenomena. Plasma instabilities are an example.[3, 4] Dynamics of other systems may be given in terms of non-linear, non-equilibrium thermo-physics. [5,6]

The AC electrolysis resonator in Puharić's device is resonant; hence the process has greater efficiency than expected had the system involved a near-equilibrium linear process.

Elizabeth A. Rauscher

1. (1-6 Superscript numbers refer to notes at the top of the following page).

REFERENCES

1. McDougall, A.. Fuel cells. MacMillan Press. 1976. p. 15.

2. Prigogine, Ilya. Physics Today. November 1972. p.23 and private communication.

3. Rauscher, Elizabeth A.. Journal of Plasma Physics. Volume 2, 517.1968.

4. Rauscher, Elizabeth A.. Bulletin of American Physical Society. Vol. 15, 1639. 1970.

5. Rauscher, Elizabeth A.. Lawrence Barkeley Laboratory memo to E. McMillan. November 18, 1980.

6. Private communication with A. Banks, Lawrence Berkeley Laboratory.

There's news and fake news and there's a massive amount of information online that is subject to restrictions and alteration. How can I be sure that this valuable information gets out and stays where anyone can access it, despite what happens to me? The answer to that is in the new concept of Blockchains. Instead of paying for hosting, which needs constant maintenance, or creating a website which can be subjected to attacks or government restrictions, the Blockchain is based on a peer-to-peer sharing concept. That means that the whole block is shared in its entirety on many people's computers. No one can change it and because of its proliferation throughout all the shared systems, it would be simply impossible to ever take it down.

Addendum:

Web3-Decentralised Website & Filestore

HYDODproject.blockchain

The www.HYDOD.com website has to be hosted, maintained and renewed every year, but the blockchain web address above is permanent and 'persistent'. That means that no matter what happens to me, it will always be there on the blockchain.

The decentralised blockchain is shared (or 'mirrored') in many locations around the world on a P2P network (Peer-to-Peer). That means it cannot be hacked, taken down or tampered with in any way. (And no, bribing won't work either, as no amount of money is worth continuing to destroy our home planet by the continued burning of fossil fuels).

The site (or 'node' as it's called) includes many files that other people can access for their own research into water-fuel technology. I've done this because I've seen too many examples of websites being hacked or disappearing altogether.

To view the blockchain website I would recommend that you use the Opera browser or for the more popular Chrome browser you can simply get a Decentralised browser extension provided by Unstoppable Domains[1], (from the Chrome Web Store[2]). It can be viewed in other browsers by changing the settings as per the instructions[3].

1. Unstoppable chrome-extension: https://unstoppabledomains.com/extension
2. Chrome Web Store: https://chrome.google.com/webstore/category/extensions
3. How to Enable Blockchain Domains in Chrome, Firefox, and Edge: https://bit.ly/3xSDRtg

Book recommendation

Fuel from Water: Energy Independence with Hydrogen – January 27, 2003, by Michael A. Peavey (Author)

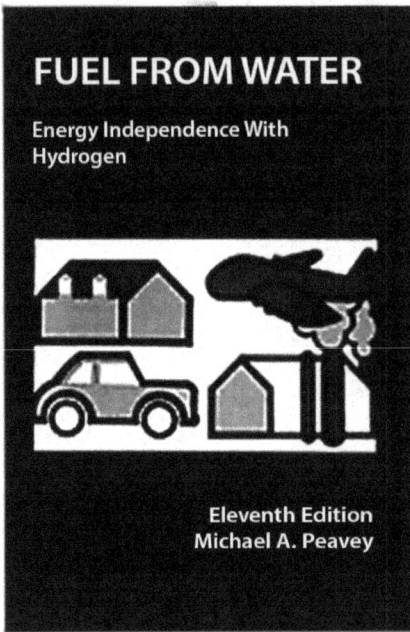

FUEL FROM WATER

Energy Independence With Hydrogen

Eleventh Edition
Michael A. Peavey

Despite what you may think, I am not getting any commission for advertising this book! I'm simply recommending it because I purchased it after hearing Stephen Meyer recommending it during his interview on Blog Talk Radio with Jas Robey. He said that he read it many times.

The book advocates hydrogen fuel as the best long-term alternative to fossil fuels and as a way to stop polluting the air and subsidising terrorists. Shows how to generate hydrogen by electrolysis, how to convert an internal combustion engine to hydrogen, and how hydrogen can be used in home appliances.

Peavey discusses the problems and benefits of using hydrogen fuel, including storage, engine modifications and stationary applications. Packed full of highly technical information, I think it is the go-to resource for anyone wanting to study this topic further.

1. Stephen Meyer interview: Stephen Meyer, Part 1: https://bit.ly/3P5gTXH
2. Energy Independence with Hydrogen Book (Amazon): https://amzn.to/3MqqyXp

Appendix 2

Website links

Alex Petty alexpetty.com

Produces an excellent website, the 'Energy Research Journal', Posts YouTube videos of his research into Stanley Meyer, Puharić, Tesla etc.

https://www.youtube.com/c/AlexPetty999

Bob Boyce

Full details, diagrams and electronics to reproduce his super-efficient 100-cell electrolyser and circuits.

https://waterpoweredcar.com/pdf.files/D9.pdf

Daniel Donatelli - Secure Supplies

Daniel Donatelli, another member of www.open-source-energy.org forum, is actively researching Puharić and Meyer's water-splitting technology and offers free circuits and 3D print files of Stan Meyer's patents.

https://danieldonatelli.wixsite.com/hydrogen-power-gas/home

https://www.youtube.com/c/SecureSuppliesLimited

https://www.thingiverse.com/search?q=Daniel+Donatelli

Effect of hydroxy (HHO) gas addition on gasoline engine performance and emissions (Article).

https://www.sciencedirect.com/science/article/pii/S1110016815001714

http://www.free-energy-info.tuks.nl/
(diy.pdf, pjkbook.pdf and hho.pdf)

Website all about cutting-edge science regarding 'free energy' and hydrogen from water, including an online store.

www.FreeFromFuel.com

YouTube videos plus a DIY manual full of design templates and loads of useful information. A paperback book and DVD to purchase, including downloadable files covering making dry cells, flame arrestors and a hydrogen welding torch, as well as hydrogen sand heaters for domestic heating.
https://www.youtube.com/watch?v=Lnm9017xcRc

www.Hydrogengarage.com

An online business selling parts to make homemade hydrogen cells, parts, circuits, etc. for the home mechanic/technical /experimenter.

Hydrogen On-Demand Website

www.HYDOD.com (See latest updates. all links will be there also).

Michael A. Peavey

"Water as Fuel" book, looks into energy independence with hydrogen. Energy Independence with Hydrogen Book (Amazon): https://amzn.to/3MqqyXp

NASA: Emissions and Total Energy Consumption of a Multi-cylinder Piston Engine Running on Gasoline and a Hydrogen-gasoline Mixture

http://ntrs.nasa.gov/archive/nasa/casi.ntrs.nasa.gov/19770016170.pdf

Patrick J. Kelly

YouTube educator and author. Produced a superb free 3,024-page eBook in 'pdf' format, 'Practical Guide to Free Energy Devices' with full details, diagrams and electronic circuits. All his work is open-source and he shares the same mission to get this technology out into the public domain. (LARGE FILE Very Slow Download! but it's also available in separate chapters)

http://vrr.dyndns.biz/Docs/OLE/FreeEnergy/PJKbook.pdf
https://www.youtube.com/user/TheEngpjk/videos

Paul Pantone Global Environmental Energy Technology (GEET)

The Global Environmental Energy Technology also known as the plasma reactor is a fuel system that was created by Paul Pantone. The GEET Fuel system is a vapour plasma unit that breaks or cracks the molecule of the fuel. This gives the engine better combustion efficiency. Let us take a look at the GEET fuel suppliers in India.

https://www.geet-pantone.com/
http://www.panacea-bocaf.org/paulpantone.htm

Petkov (Valentin Petkov /valyonpz/) - Electronics engineer - YouTuber

As a member of open-source-energy.org, he has contributed a great deal to the research into the use of water as a fuel source. An active YouTuber. posting many videos of his research, openly sharing his analytical conclusions.
https://www.youtube.com/c/valyonpz

Ritalie - 'Radiant energy'

Ritalie's circuits harness 'back EMF' (electromotive force) He calls it 'radiant energy'. For water-fuel cells and charging batteries - "auto-tune" self-resonating, solid-state circuits. His website sells circuits, eBooks and parts.

https://www.youtube.com/c/Ritalie

https://ritalie.com/store

Robert Murray Smith (YouTube Graphite dry cells etc).

YouTube channel exploring practical scientific experiments. including many relevant to this book.

https://www.youtube.com/c/RobertMurraySmith

Russ Gries

Russ Gries runs a very lively forum and highly informative YouTube channel about all things to do with using water as a fuel and is obviously keen to share all his research.

www.rwgresearch.com www.open-source-energy.org

https://www.youtube.com/user/rwg42985

Water Spark Plugs (Plasma) by Panacea-BOCAF On-Line University
http://www.panaceatech.org/Water%20Spark%20Plug.pdf

The Wonn YouTube
Research and development of alternative fuel and energy sources.

https://www.youtube.com/user/thewonn

YouTube videos

Editors note: All these links are available on the www.hydod.com website for your convenience. You can also search YouTube by their names. Please let me know if you find any that no longer work, thanks.

'3 Balloons' (Filled with: 1. Hydrogen, 2. Oxygen 3. Hydrogen and Oxygen) College of Chemistry, University of California, Berkeley

https://www.youtube.com/watch?v=a6qGIMqDKwA

Best Demonstration of Resonance -MIT professor demonstrates how glass breaks due to forced resonance

https://www.youtube.com/watch?v=pyBGSMxPSG0

Generator running on 100% hydrogen HHO

https://www.youtube.com/watch?v=uex7ItMSZ5Y&t=261s

Generator running on HHO testing

https://www.youtube.com/watch?v=mhbVzRfYpr0&t=471s

HHO GENERATOR INSTALLATION ON DODGE RAM 1500 V8

https://www.youtube.com/watch?v=ECNO_FFk7Hw

HHO Generator - Water to Fuel Converter

https://www.youtube.com/watch?v=cqjn3mup1So

Run your car on water. This guy does just that!

https://www.youtube.com/watch?v=wjeM2IBhtlc&t=82s

Stan Meyer - 'It runs on water' (The UK's Channel 4 documentary 1980)

https://www.youtube.com/watch?v=t98UBY3GhhI

Stan Meyers Water Car

https://www.youtube.com/watch?v=GFIlXaABU54

Stanley Meyer's water car's first run

https://www.youtube.com/watch?v=Z5afwEcZ3Ok

Stan Meyer explains the (Water Fuel Technology) on the dune buggy

https://www.youtube.com/watch?v=EJ-zb0CmGag&t=209s

Stanley Meyer Water Fuel Cell hard copy

https://www.youtube.com/watch?v=MK9HDC2fFTk&t=21s

Video Chemistry Org (Excellent intro to hydrolysis).

https://www.youtube.com/watch?v=dRtSjJCKkIo

Water Fuel-Cell Inventor Murdered by Government.

https://www.youtube.com/watch?v=gIAZFU2rAQE&t=57s

Water Fueled Spark Plugs Nano Bubble Water Fuel Ignition System

https://www.youtube.com/watch?v=WF619fsTGbM

Patents

These patents, all particularly relevant to the book's message, are all in the public domain and currently freely available to anyone (e.g. online from Google patents). Some have expired and some apply only to the USA.

(Or search for the number, name or title, on https://patents.google.com/)

Puharić's Patents

Control & Driver Circuits for a Hydrogen Gas Fuel Producing Cell
WO 92/07861: https://bit.ly/3GDPTL7

Controlled Process for the Production of Thermal Energy from Gases
USP # 4,826,581: https://bit.ly/3NPF7DQ

Gas Electrical Hydrogen Generator
USP # 4,613,304 : https://bit.ly/3z9wakQ

Gas Generator Voltage Control Circuit
USP # 4,798,661 : https://bit.ly/3MbnkWy

Hydrogen Gas Burner
USP # 4,421,474: https://bit.ly/3a91Kop

Hydrogen Gas Injector System for Internal Combustion Engine
USP # 4,389,981: https://bit.ly/38CbdEg

Method and Apparatus For Splitting Water Molecules
USP # 4,394,230: https://bit.ly/3M9owtJ

Start-up/Shut-down for a Hydrogen Gas Burner
USP # 4,465,455: https://bit.ly/3PWYugo

Means for aiding hearing by electrical stimulation of the facial nerve system
Patent US3170993A: https://bit.ly/3x3gEV6

Stephen Meyer's Patents

Hydroxyl Filling Station
https://patents.google.com/patent/US20050246059A1/en

Stanley Meyer's patents
Below is a partial list of Meyer's patents that are now in the public domain

Controlled process for the production of thermal energy from gasses and apparatus useful therefore - expired 5/02/2006
https://patents.google.com/patent/US4826581

Electrical pulse generator - expired 9/23/2003
https://patents.google.com/patent/US4613779

Gas electrical hydrogen generator - expired 9/23/2003
https://patents.google.com/patent/US4613304

Gas generator voltage control circuit - expired 1/17/2006
https://patents.google.com/patent/US4798661

Hydrogen/air and non-cumbustible gas mixing combustion system
https://patents.google.com/patent/CA1227094A

Hydrogen airdation injection system
https://patents.google.com/patent/EP0122472A2/sv

Hydrogen gas burner - expired 8/25/2002
https://patents.google.com/patent/US4421474

Hydrogen gas fuel and management system for an internal combustion engine utilizing hydrogen gas fuel - expired 3/15/2011
https://patents.google.com/patent/US5293857

Hydrogen gas injector system for internal combustion engine - expired 2/17/2002
https://patents.google.com/patent/US4389981

Method for the Production of a Fuel Gas - expired 8/05/2007
https://patents.google.com/patent/US4936961

Process and apparatus for the production of fuel gas and the enhanced release of thermal energy from such gas - expired 9/22/2009
https://patents.google.com/patent/US5149407

Resonant cavity hydrogen generator that operates with a pulsed voltage electrical potential
https://patents.google.com/patent/CA1234773A

Start-up/shut-down for a hydrogen gas burner - expired 9/24/2002
https://patents.google.com/patent/US4465455

Water Fuel Injection System
https://patents.google.com/patent/CA2067735A1/en?

Nikola Tesla's bifilar coil (United States patent 512,340 of 1894
https://patents.google.com/patent/US512340A/en?

The Francisco Pacheco Bi-Polar Auto Electrolytic Hydrogen Generator
Article: https://fuel-efficient-vehicles.org/energy-news/?page_id=926
https://patents.google.com/patent/US5089107A/en?oq=5089107

The Francisco Pacheco Gas-operated internal combustion engine
https://patents.google.com/patent/US3648668A/en?

Best Demonstration of Resonance -MIT professor demonstrates how glass breaks due to forced resonance
https://www.youtube.com/watch?v=pyBGSMxPSG0

Wine glass resonance in slow motion
https://www.youtube.com/watch?v=BE827gwnnk4

Breaking a wine glass using resonance
https://www.youtube.com/watch?v=17tqXgvCN0E